The mathematical work of Charles Babbage

D0068807

CHARLES BABBAGE

The mathematical work of
Charles Babbage

J.M.DUBBEY

Head of Department of Mathematical Sciences and Computing
Polytechnic of the South Bank, London

Fairleigh Dickinson
University Library

Teaneck, New Jersey

CAMBRIDGE UNIVERSITY PRESS

CAMBRIDGE

LONDON · NEW YORK · MELBOURNE

340070

Published by the Syndics of the Cambridge University Press
The Pitt Building, Trumpington Street, Cambridge CB2 1RP
Bentley House, 200 Euston Road, London NW1 2DB ·
32 East 57th Street, New York, NY 10022, USA
296 Beaconsfield Parade, Middle Park, Melbourne 3206, Australia

© Cambridge University Press 1978

First published 1978

Printed in Great Britain at the University Press, Cambridge

Library of Congress Cataloguing in Publication Data

Dubbey, John Michael, 1934 –
The mathematical work of Charles Babbage.

Includes index.

1. Mathematics – Great Britain – History. 2. Computers – History.

3. Babbage, Charles, 1792–1871. I. Title. QA27.G7D8   519.4'092'4
77–71409

ISBN 0 521 21649 4

QA
27
·G7
D8

CONTENTS

PREFACE

This book is a critique of the mathematical work of Charles Babbage. In it I have attempted to demonstrate a continuous development from his early work in notational reform with the Analytical Society to his ideas and construction of the difference and analytical engines. I have discussed his achievements within the context of the mathematical climate in which he lived, and the range of fundamental thinking and inventiveness he applied to non-mathematical concepts. He had ideas well ahead of his time that were misunderstood and neglected by his contemporaries, but which place him as a significant figure in the development of modern mathematics.

One continuous thread in Babbage's personal development was the importance he attached to notation, and in advancing from his powerful advocacy of the differential notation for the calculus to the need for a mechanical notation to describe the working parts of his engines, he laid down rules and even experimented with mathematical symbolism which to him provided the unity between the concepts of generality and simplicity. It has been necessary, in order to avoid excessive problems in printing, not to reproduce all of these notational devices precisely; in particular, standard subscripts and superscripts have been used in some places instead of Babbage's vertical notation. One typical example from chapter 5, is that the original $\overset{p}{\underset{n}{S}}$ is printed as S_n^p, but this clearly makes no difference to the content.

Finally I would like to thank those who have helped in this work. Dr A. V. Armitage, Dr H. B. Calvert, Dr D. T. Whiteside and Mrs E. Butler made many valuable suggestions, and Mrs W. Whitaker typed accurately through the somewhat intricate notation. I am most grateful to the staff of the Cambridge University Press for their efforts to make the manuscript more coherent and presentable. Any errors are my own responsibility.

1

Introduction

Charles Babbage is an almost legendary figure of the Victorian era, yet relatively little is known about him. No authoritative account of his life and work has yet been published. In the absence of accurate knowledge, he is misrepresented as the eccentric genius, inventing computers which he never completed and quarrelling with almost everyone, especially the organ-grinders. Certainly Babbage was a man of highly individual talent, applying his ability with great success to a variety of subjects from economics to ballet, from deciphering to life insurance, from tool-making to astronomy. Indeed it is hard to think of any field of knowledge in which he did not excel, excepting possibly classics, for which he admitted a dislike. As far as his personal life is concerned, there is abundant, scarcely touched material available for studies of his exceptional personality, and when this is thoroughly examined it will almost certainly be discovered that he was a far different person from the one represented by popular misconception.

Primarily, Charles Babbage was a mathematician. In spite of the great variety of interests in other spheres, together with a considerable amount of family and social commitment, there is no doubt that he devoted himself essentially to a study of pure mathematics during the early years of his working life. In these years, as will be shown, his productive work was most original in content, exerting a strong influence on the course of British mathematics. It was recognised and esteemed by some of the greatest contemporary Continental mathematicians and contained many ideas, the value of which was not acknowledged till many years later.

At the age of thirty, Babbage embarked on his great life work:

1

the invention and manufacture of first the 'difference engine' and then the 'analytical engine'. This proved to be a task of such magnitude that it occupied him for the remaining fifty years of his life, but it was undoubtedly his major contribution to modern knowledge. During these years, he appeared to have abandoned his studies of pure mathematics, as far as published work was concerned. Nevertheless he always sought to provide a mathematical approach to each of the many and varied problems which he tackled. Whether discussing miracles, pin-making, postal services, geology, economics, politics or even his private life, he always attempted to formulate the problem in as mathematical a way as possible. It was his misfortune to encounter an age that was not so logical as himself. Mathematics was his first love, and it is largely because of his work that the world today, for better or worse, is much more mathematically aligned than it was a hundred years ago.

For these reasons I believe that it is not possible to assess the importance of Charles Babbage without an appreciation of his mathematical work. Hitherto, this has never been attempted. Most of his published mathematical work has not been read, let alone considered in relation to his subsequent life, and it is this defect which I propose to try to remedy. By attempting a critique of the whole of Babbage's mathematical methods and output, I hope to secure a necessary bridge-head from which the rest of his work can be properly analysed and appreciated.

Various accounts of Babbage's life and work are available. His own autobiography is *Passages from the Life of a Philosopher*, London, 1864. This is a most entertaining and interesting work. An unsystematic biography, it contains reminiscences and anecdotes, the historical accuracy of which cannot always be guaranteed. The reader receives a variety of information about the author, but learns nothing of such normally important biographical details as his place and date of birth, his wife and children.

The story of Babbage's life has recently been told in Maboth Moseley's work (the title of which indicates her general assessment of his character), *Irascible Genius*, London, 1964. This is the best biographical account of Babbage available, but contains little scientific appraisal.

There is also a long, unpublished and undated biography by Babbage's friend, Harry Wilmot Buxton, which is at present kept in a cowhide trunk containing several other relics of Babbage in the Museum of the History of Science, Oxford. The full title of this work is

> Memoir of the life and labours of the late Charles Babbage Esq. F.R.S. formerly Lucasian Professor of Mathematics in the University of Cambridge. Comprising a description and historical account of his Analytical and Difference engines derived principally from his posthumous M.S.S. and papers by Harry Wilmot Buxton Esq., F.R.A.S.

The reference to the 'late Charles Babbage' in the title, and the fact that much of the text reads like an extended obituary notice, indicate that the work was compiled shortly after his death in 1871. It is written in a leisurely style and occupies over a thousand pages. The script is written on even-numbered pages only, with a corresponding blank odd-numbered sheet to each written one, intended presumably for correction and amendment. The author has made little use of these blank sheets, indicating that he did not make much progress in revising the text.

The book contains some biographical details of Babbage's early life which are probably more accurate than the ones from the *Passages from the Life of a Philosopher*. There is an account of his life at Cambridge, accompanied by a very long quotation from one of Babbage's unpublished works: 'The History of the Origin and Progress of the Calculus of Functions during the years 1809, 1810 . . . 1817.' As the title indicates, the bulk of the book is a very full account of the difference and analytical engines. It would appear that Buxton obtained his material for this subject from manuscripts that I found were still available when writing the eighth chapter of this book. There is also an interesting assessment of the work and character of Babbage, parts of which will be quoted in our final chapter. Although Buxton's tribute to Babbage is written in the most glowing terms, even he could have had little idea of Babbage's importance when viewed from a twentieth-century perspective.

Babbage produced volumes, some consisting entirely of diagrams, about his calculating engines; but relatively little was published. More detailed accounts of these machines were presented, with Babbage's full approval, by Dionysius Lardner on the difference engine, and by L. F. Menabrea and Lady Lovelace on the analytical engine. Most of the relevant articles, together with reports from the Royal Society and the British Association,

were collected and published by Babbage's son, Henry Prevost
Babbage, in *Babbage's Calculating Engines*, London, 1889. There
is also a more recent book containing a collection of most of the
published accounts of the engines edited by P. and E.
Morrison, *Charles Babbage and his Calculating Engines*, New York, 1961.

Amongst shorter articles on Babbage must be mentioned the
entry in the *Dictionary of National Biography* by J. A. Fuller
Maitland; L. H. Dudley Buxton's 'Charles Babbage and his difference engines', *Transactions of the Newcomen Society*, 1934, **14**,
43–65, based on his grandfather's unpublished book; and S. H.
Hollingdale's 'Charles Babbage and Lady Lovelace – two 19th
century mathematicians', *Bulletin of the Institute of Mathematics and
its Applications*, 1966, **2**, 2–15.

Most modern books on computers contain a chapter of historical introduction in which acknowledgement to Babbage is made,
the fullest of these accounts and the warmest of tributes appearing in B. V. Bowden's *Faster Than Thought*, London, 1953.

Babbage himself lists eighty items of his published work in an
appendix to *Passages from the Life of a Philosopher* and at least eight
of these are full-length books. There are, in addition, numerous
unpublished works, including sixteen bulky volumes of correspondence in the British Museum Manuscripts Room. For the
purposes of this book I have been particularly glad to discover
two complete, unpublished mathematical works, the one in
Oxford, already referred to, and 'The Philosophy of Analysis' in
the British Museum Collection.

There has been some doubt as to when and where Charles
Babbage was born. This arises mostly from his own indifference
to the subject. H. W. Buxton gives 26 December 1791 as the date;
L. H. D. Buxton 26 December 1792. The article in the *Dictionary
of National Biography* states that he was born near Teignmouth on
26 December 1792, and P. and E. Morrison more cautiously
suggest somewhere in Devonshire in 1792. M. Moseley quotes
Babbage as saying on a certain occasion that he was born in
London in 1792; she disagrees, stating categorically that he was
born in Totnes on 26 December 1791.

The Times in its obituary notice of 23 October 1871 reported
that 'Little is known of Mr. B's parentage and early youth except
that he was born on the 26th December 1792.' A week later, the
Rev. T. H. Hollier, a nephew of Babbage, wrote in to correct this
statement to 26 December 1791.

The Marylebone Society tried unsuccessfully during the 1960s to resolve this question, in order to acquire the right information for a memorial plaque, but the solution was discovered by A. Hyman in 1975. He found a baptismal entry in the Parish Register of St Mary's, Newington, London: '6th January 1792 Charles s. of Benjamin and Betsy Plumleigh Babbage.'

Since 26 December seems to be the undisputed birthday, the correct year appears to be 1791, and Hyman suggests the most probable place of birth as the family residence of 44 Crosby Row, Walworth Road, very near to the Elephant & Castle, London.

I am indebted to H. W. Buxton for other details of Babbage's early life, and some of these are corroborated in *Passages from the Life of a Philosopher*. Charles was one of four children born to Benjamin Babbage, a member of the firm of Praed, Mackworth and Babbage (the London bankers), and Betsy Plumleigh Teape. His two brothers died in infancy, but his sister lived and survived him (*ibid*. p. 66).

He was a very weak and sickly child during his early years, and his parents were advised not to worry him with any kind of education. He grew stronger in adolescence, and, according to H. W. Buxton 'became finally exempt from those real or imaginary ills which had hitherto held him in thraldom' (*op. cit.*, p. 68). Babbage continues in *Passages from the Life of a Philosopher* (p. 10):

> Having suffered in health at the age of five years, and again at that of ten by violent fevers, from which I was with difficulty saved, I was sent into Devonshire and placed under the care of a clergyman (who kept a school at Alphington, near Exeter), with instructions to attend to my health; but, not to press too much knowledge upon me: a mission which he faithfully accomplished.

According to Buxton, he was then removed to the academy of the Rev. Stephen Freeman of Forty Hill, Enfield, Middlesex. This school consisted of thirty boys, one of Babbages's earliest friends being Frederick (later Captain) Marryat. His formal education began at this stage. He showed at once an indifference to the classics, but was strongly influenced by John Ward's *Young Mathematicians' Guide*, London, 1707.

After an unspecified number of years at Enfield, Babbage resided at home under the guidance of an Oxford tutor. In his now considerable spare time, he began to read as many mathematical books as he could find.

Amongst these were Humphry Ditton's 'Fluxions', of which I
could make nothing; Madame Agnesis' 'Analytical Institutions',
from which I acquired some knowledge; Woodhouse's 'Principles of Analytical Calculation', from which I learned the notation of Leibnitz; and Lagrange's 'Theorie des Fonctions'. I
possessed also the Fluxions of Maclaurin and of Simpson. (*Ibid.*
p. 26.)

According to his unpublished work on the calculus of functions, he began his researches into this subject at about this time,
late 1809.

He entered Trinity College, Cambridge, in October, 1810, but
migrated to Peterhouse where he took his B.A. in 1814, graduating as M.A. in 1817. His time at Cambridge was most fruitful
mathematically. He soon found that what he had learnt from
private study was more than that required for examination purposes, and consequently he ignored the formal tuition. He
formed the Analytical Society together with J. F. W. Herschel,
George Peacock and others, for the conducting of mathematical
discussion and research. The work of this Society had a very
beneficial effect on British mathematics, which had severely
declined by the end of the eighteenth century.

At this time, Babbage and Herschel wrote the book *Memoirs of
the Analytical Society*, Cambridge, 1813, a work which indicated
the depth of their reading and the originality of their research.
Babbage also had his two long essays on the calculus of functions
published in the *Philosophical Transactions* of 1815 and 1816.

On leaving Cambridge in 1815, Babbage resided in London at
5, Devonshire Street. He was elected Fellow of the Royal Society
in 1816 and Fellow of the Royal Society of Edinburgh in 1820. He
helped to found the Royal Astronomical Society in 1820, being
Secretary for 1820–24 and later Vice-President, Foreign Secretary and Member of the Council.

Little is known about his family life. He married Georgina
Whitmore in 1814 and they had eight children, four of whom
survived childhood. 1827 was a year of personal disaster in which
Babbage lost his father, his wife and two of his children, his own
health breaking down. He never married again.

In about 1821, be began work on his engines, a task which
occupied most of the rest of his working life, and involved him in
controversies with the Government and others. The details of this
part of Babbage's life are related in chapter 8.

Becoming increasingly critical about the state of scientific organisation in this country, he wrote a polemical book, *Reflections on the Decline of Science in England*, London, 1830, expressing great dissatisfaction with the Royal Society. He helped to found the British Association in 1831, being a Trustee from 1832–38. He also originated the Statistical Section of this Association at the Cambridge meeting in 1833 and helped to found the Royal Statistical Society in 1834, being Chairman in 1835.

In 1828 he was appointed Lucasian Professor of Mathematics at Cambridge and held this position for eleven years without, it would appear, ever residing in Cambridge or giving a lecture.

He moved to 1 Dorset Street, Portman Square, London, in 1827 and lived there until his death on 18 October 1871.

He was honoured by many foreign learned societies, especially in Italy where his work was greatly esteemed. To quote from the title page of *Passages from the Life of a Philosopher*, he describes himself as

> Charles Babbage, Esq., M.A.,
> F.R.S., F.R.S.E., F.R.A.S., F.Stat.S., Hon.M.R.I.A., M.C.P.S., COMMANDER OF THE ITALIAN ORDER OF ST. MAURICE AND ST. LAZARUS, Inst. Imp. (Acad. Moral.) Paris Corr., Acad. Amer. Art et Sc. Boston, Reg. Occon. Boruss., Phys. Hist. Nat. Genev., Acad. Reg. Monac., Hafn., Mussil., et Divion, Socius, Acad. Imp. et Reg. Petrop., Neap., Brux., Patar., Georg. Floren, Lyncii Rom., Naut., Phiomath, Paris, Soc. Corr. Etc.

After this brief sketch of the major reference points in Babbage's life, I now turn to the contents of the rest of this book.

Chapter 2 is an attempt to give a background to Babbage's mathematical life by a consideration of the state of the subject in this country at the time of his research. It is now almost banal to say that British mathematics had reached a very low mark by the beginning of the nineteenth century. Hardly any detailed study of this period has been presented, and I have attempted here to give the full picture of what was happening. The result is a study at various levels: in school, university, published books and research papers. More information is obtained by looking at general attitudes to the subject and attempts to reform. An extraordinary situation is revealed. It appears that mathematics was so highly esteemed at Cambridge that it was the only subject for all students at undergraduate level, and yet it was, in comparison with what was happening on the Continent, virtually neglected for research, industrial or military purposes. Possibly the reason for this paradox is that there is no better way to place people arbitrarily in order of merit than to give them an examination in mathematics, and this was the sole object of the exercise at Cambridge.

Against this background Babbage had to teach himself all the mathematical knowledge he needed. He was very fortunate to have such friends as Herschel, Peacock, Maule, Bromhead and others of a similar mind to his own, but his ability proved to be superior to any of these. By his own discoveries and his general encouragement of mathematical learning, he helped to herald a new era in British mathematics.

The third chapter looks in detail at the formation and work of the Analytical Society, in which Babbage played a considerable part. In particular I discuss the campaign to introduce the differential notation used by the Continental followers of Leibniz to the compatriots of Newton who had generally refused to use it for over a century.

There can be little doubt that Babbage's major mathematical work was in the calculus of functions, and this is studied in some detail in chapter 4. It is shown how, particularly in the two long essays written for the *Philosophical Transactions* of the Royal Society, he took a branch of mathematics barely considered by his predecessors and transformed it into a systematic calculus, the analysis containing some very original stratagems and devices. Justice, it is claimed, has not really been done to his achievement in this field, and I have made an attempt to rectify this.

It was a great pleasure to discover his unpublished book, 'The Philosophy of Analysis', which is examined in chapter 5. There are many interesting items in this work, the most notable being the author's views on algebra, which were far in advance of his time and which anticipated many of the early theories of so-called modern algebra. His analysis of the game of noughts-and-crosses, in which he deduces something of a stochastic process, is also of interest.

Babbage was a versatile mathematician, and chapter 6 surveys some of his other mathematical interests, ranging over geometry, the theory of numbers, probability, and finite and infinite series; but it is not claimed that his mathematical thought was very powerful in these directions.

Chapter 7 is concerned with Babbage's views on notation. Throughout his career he emphasised the vital importance of a good working symbolism. At Cambridge he crusaded successfully for the reformation of notation in the differential calculus. Later he published his views at greater length, not only devising a set of rules that all mathematical notations were to follow but even constructing a kind of notational calculus. We can see continuity from the mathematical to the less mathematical part of his life, when he later devised, as an essential feature of his drawings for the engines, a workable convention which he described as the 'mechanical notation'.

It would not seem proper to discuss Babbage's mathematical work without a consideration of his pioneering work in the computer field. Strictly speaking, the construction of a computer depends on logical rather than specifically mathematical principles, but there is an obvious intimate connection between computers and mathematics. His work on the engines also illustrates the fact that he never abandoned mathematics, but continually used the type of thinking involved in areas that were not so obviously mathematical.

To consider Babbage's work on computers at all, necessarily involves a lengthy discussion, as he strived for practically fifty years in this field. Chapter 8 describes the origin, principle and attempted construction of the difference and analytical engines, without becoming too involved in the mechanisms, while considering all the difficulties that Babbage encountered. His major obstacles were lack of adequate finance, wavering support by the Governments of his day, misunderstanding by the general public, the paucity of precision engineering, and above all his own mind, which so raced ahead of what it was possible to achieve practically that none of these major projects were ever completed.

Finally, I have tried in the last chapter to survey Babbage's major mathematical achievements as seen from the twentieth century, and to assess the influence and importance of his work in the context of the present age.

2

British mathematics 1800–30

The period between 1800 and 1830 was not a brilliant one for British mathematics by any standards. The decline which had begun early in the eighteenth century was continued. Very little original work of any great merit was accomplished and it is difficult to name a single British mathematician of this time whose work is still remembered. By contrast, these were years of great advance on the Continent. Gauss was producing his best work, while in France, Lagrange, Laplace, Legendre, Cauchy, Fourier, Poisson, Monge and Poncelet, to mention a few prominent names, were all active. Who were the contemporary British mathematicians who compared with these? The same question was asked in 1830 by A. B. Granville, but he had a ready answer: 'Ivory, Woodhouse, Morgan, Herschel, Babbage, Kater, Christie, Barlow, Baily, Gompertz, Whewell, Allman, Peacock, Lubbock, Bromhead and Groombridge.'[1] This was a list which evidently satisfied its compiler, but bears no comparison with the Continental mathematicians.

However, the period is interesting as a time of constructive introspection, when mathematical reformers looked for the causes of the sickness and tried to cure them, a necessary preliminary to the great advances made in mid-century by such mathematicians as Cayley, Sylvester, Maxwell, Hamilton and Boole.

There are several reasons for the British decline, and most of these relate to the giant figure of British mathematics, Sir Isaac Newton. His retirement from active research at the beginning of the eighteenth century created a vacuum which mathematicians of the calibre of Cotes, De Moivre, Taylor and Maclaurin could only partially fill. No-one was able to develop Newton's great discoveries in gravitation and the calculus to any significant degree; and as the century progressed, the decline in talent became more apparent. It probably reached its lowest level at the start of the nineteenth century. A writer in the *Edinburgh Review* of 1807 had a better idea of the true situation in Britain than Granville. In the course of a lengthy review[2] of P. S. Laplace's *Traité de Mécanique Céleste* he turned his attention to the contrast between Continental and British mathematics and remarked that, in the course of discussing the progress of mathematics and theoretical astronomy during the previous seventy years, he had not been able to mention the contributions of any British mathematician. There seemed to be a complete lack of knowledge of the work being done on the Continent. According to the reviewer,

> At the moment when we now write, the treatises of Maclaurin and Simpson, are the best which we have on the fluxionary calculus, though such a vast multitude of improvements have been made by the foreign mathematicians ... A man may be perfectly acquainted with every thing on mathematical learning that has been written in this country, and may yet find himself stopped at the first page of the works of Euler or D'Alembert.[3]

With reference to the *Mécanique Céleste* he remarks that there were probably not more than a dozen men in Britain who could read the work, let alone add to any of its conclusions.

However, there was one Englishman at least who could read it. The Rev. John Toplis translated parts of the *Traité de Mécanique Céleste*, supplying his own notes. A year after graduating as M.A. from Cambridge in 1804, he wrote his own tirade against the then current state of British mathematics, in the *Philosophical Magazine*, a journal that published many scientific but very few mathematical articles. 'It is a subject of wonder and regret to many, that this island, after having astonished Europe by the most glorious display of talents in mathematics and the sciences dependent upon them, should suddenly suffer its ardour to cool, and almost entirely to neglect those studies in which it infinitely excelled all other nations.'[4]

The rift between British mathematicians and their Continental contemporaries undoubtedly went back to the utterly barren controversy over the order of priority in the invention of the calculus between the supporters of Newton and of Leibniz. The bitterness aroused by this dispute resulted in a general lack of co-operation and friendship between mathematicians on either side of the Channel which lasted for nearly a hundred years. This loss of contact only emphasised the inferiority of the British work, and the depth of partisan hostility in this private war between otherwise rational men is borne out by the curious fact that the first sign of a truce, of any admiration for or co-operation with French mathematicians, occurred at the time of the war with Napoleon. The Rev. Baden Powell, writing of the controversy shortly after its resolution, described it as

> One of the most bitter and keenly contested, perhaps, which any disputed topics (scarcely excepting those of theology) have ever called forth – it led to an almost total cessation of that mutual interchange of information and opinions on scientific subjects, which is always so highly beneficial to both the parties engaged. Long after the actual question of the original controversy had fairly worn itself out, the jealousy which was felt, and the line of separation which had been drawn between the British and continental mathematicians, were maintained in their full force, and produced the most pernicious effects on science. Each party became the exclusive supporters of the system taught by the two great luminaries of their respective countries.[5]

This separation meant that British mathematicians, out of a sense of loyalty to their national hero (as Robert Woodhouse put it: 'My more than rational reverence for Newton'[6]), used only his methods and notation. They obstinately refused to accept the more analytical and suggestive approach of Leibniz, used so successfully by his followers on the Continent. Newton had constrained himself to working within a framework of Euclidean geometry and attempted to relate the calculus to dynamical or fluxionary principles, with their awkward 'dot' notation. He succeeded in spite of these self-imposed handicaps but his less able followers, in the letter if not the spirit, were quite unable to overcome such obstacles. Baden Powell remarks:

> The differences in name and notation between the two methods, though in itself a trivial circumstance, was yet far from unimportant in some of the consequences which may be fairly traced to it. It tended in some measure to foster and increase the dissension between the two schools, and their ignorance of each other's researches; while the diversity itself between the two

methods, though in reality little more than nominal, became also a topic of no small dispute and controversy.[5]

A further disadvantage for British mathematics was that it was torn by internal dissent in the middle of the eighteenth century by an attack on the whole basis of Newton's calculus. The philosopher George Berkeley had written a tract called *The Analyst*, London, 1734, in which he pointed out the thoroughly illogical and non-mathematical basis of the calculus as Newton had left it. In fairness it should be said that up to this time Newton was one of the few mathematicians to have acknowledged logical difficulties in the fundamental principles of the differential calculus, and to have actually made several attempts to resolve them. Berkeley showed in a convincing way that he had not been successful. *The Analyst* was fiercely attacked by supporters of Newton, and many books and tracts were written in the next ten years, nearly all attempting to refute Berkeley and succeeding only in demonstrating both the ignorance of these writers (mostly professors of mathematics) and their own total misunderstanding of the first principles of calculus. These misconceptions of the calculus continued until the following century, and Woodhouse, writing in 1803, urged all mathematicians to read *The Analyst*, since 'If it does not prove the cure of prejudice, it will be at least the punishment.'[7]

This is the general background to the period under study. It will be found that British mathematicians were riddled by irrational prejudice, unhealthy conservatism in choice of method and notation, cut off from their Continental contemporaries and uneasy about unsettled internal disputes.

There was also the feeling that mathematics had little intrinsic importance. In France, the best mathematicians had been given prominent positions in the State and every encouragement to pursue their researches both for the immediate national need and the general progress of the subject. The British attitude was very casual. This was typified in the preface to Ludlam's *The Rudiments of Mathematics*, which was then one of the most popular university textbooks: 'It is certainly with good reason that young scholars in our Universities are put upon the Study of Mathematics. But surely it is not so much for their intrinsic value, as their use in qualifying young men to reason justly on more important subjects.'[8] This Platonic idea of mathematical studies being a means to an end for the training of the mind for 'higher' things was prominent, as the teaching given at universities during this time shows.

The writer in the *Edinburgh Review*, continuing his lament over the dearth of mathematicians in England, asserts that the talent is there, but diverted into trivial pursuits:

> The Ladies' Diary, with several other periodical and popular publications of the same kind, are the best proofs of this assertion. In these, many curious problems, not of the highest order indeed, but still having a considerable degree of difficulty, and far beyond the mere elements of science, are often to be met with; and the great number of ingenious men who take a share in proposing and answering these questions, whom one has never heard of any where else, is not a little surprising. Nothing of the same kind, we believe, is to be found in any other country.[9]

This mis-direction of ability was also referred to by Baden Powell who added:

> It is not twenty years since we have begun to perceive that we were far behind all the rest of Europe in these sciences; not from want of abundance of first-rate talent, but from a misapplication of that talent to unworthy objects, or at least to such as were not of a nature calculated to lead to any great advance in the state of knowledge.[10]

The feeling that mathematics was only a charade, a useful exercise for sharpening the wits, prevailed with few exceptions throughout this period. This attitude was reflected in the educational system, where the subject was not in most cases taken very seriously. The Clarendon Report of 1864[11] indicated that mathematics teaching in public schools was almost non-existent during the first thirty years of the nineteenth century. Of the nine public schools examined in detail, the following three reports will suffice to indicate the findings of HM Commissioners.

> Before the year 1836 there appears to have been no mathematical teaching of any kind at Eton.[12]

> The study of Mathematics was first made compulsory at Harrow in 1837. Before that time it had been voluntary; the present Senior Mathematical Master, Mr. Marillier, gave private lessons to such boys as desired it. He had himself been at the school since 1819. When he came there mathematical

instruction could only be obtained from a Writing Master (who was then very old), except that the boys in the Sixth Form read Euclid once a week with the Head Master.[13]

The report on Rugby was candid about the quality of the teaching available in the early part of the century:

> Arithmetic became a part of the instruction given at Rugby in the year 1780.
>
> The method of providing Mathematical Masters for the school has varied greatly during the course of the present century. In the decennial period between 1820 and 1830, two Masters, neither of whom had taken degrees in either University, taught the whole school.[14]

It added that these two unfortunate men commanded little respect in the classroom, among their fellow staff, or in the town. The report also stated that 'Natural Philosophy became a subject of instruction at Rugby in the year 1849.'[15]

It is difficult to assess the situation in other types of school at this time. There is the much later Newcastle Commission of 1861, appointed to enquire into the state of popular education in England. This contains a most interesting statistical analysis of subjects taught in schools, and, despite its reference to a time thirty years after the period under consideration, will nonetheless

Subjects	Public week-day schools	Private week-day schools	Evening schools
Religious	93.3	71.7	63.2
Reading	95.1	93.5	85.1
Writing	78.1	43.2	85.2
Arithmetic	69.3	33.8	73.7
Needlework	75.8	73.8	17.0
Other Industrial Work	3.8	3.38	—
Mechanics	.6	1.29	.7
Algebra	.8	1.35	.7
Euclid	.8	1.15	.6
Elements of Physical Science	3.1	1.84	1.2
Music from Notes	8.6	3.1	1.9
Drawing	10.8	2.98	1.9

be useful for informative and comparative purposes. The Commission obtained evidence from 1824 public week-day schools, 3495 private week-day schools and 681 evening schools. The findings included the above tables[16] which indicate the percentage of scholars taking a particular subject.

It is most likely that these figures represent upper bounds to any that might have been produced in 1800–30. They indicate that, even at this later period, arithmetic was taught only at roughly half of the schools and that any 'higher' mathematics was almost completely ignored. It is interesting to note that, in a further breakdown of these figures among the sexes, no females were taught mechanics, algebra or Euclid at any institution. The tables suggest that where any sort of mathematics was taught at school it was almost invariably arithmetic. It was not particularly good arithmetic either, as an example from one of the most popular textbooks, F. Walkingame's *The Tutor's Assistant*, London, 1751, used well into the nineteenth century, will indicate. To explain compound interest, the usual symbols A, P, T and R are defined and two tables supplied, the first to convert percentages into decimals, and the second to give the amount after any number of years at five and six per cent. The text continues:

'when P, T, R, are given to find A
RULE $p \times r^t = A$
EXAMPLES'[17]

There is no explanation of the principle. Just the rule and examples of how to put numbers into the formula and perform the necessary arithmetic are given. Even the variables are changed from capital letters to small without explanation. The book gives no incentive at all to mathematical thinking.

Mathematics was an integral part of the studies at most British universities, and the situation there was considerably better than at the schools. However, the writer in the *Edinburgh Review* sees the two major universities as the chief culprits responsible for the mathematical decline:

> It is chiefly in the public institutions of England that we are to seek for the cause of the deficiency here referred to, and particularly in the two great centres from which knowledge is supposed to radiate over all the rest of the island. In one of these, where the dictates of Aristotle are still listened to as infallible decrees, and where the infancy of science is mistaken

for its maturity, the mathematical sciences have never flourished; and the scholar has no means of advancing beyond the mere elements of geometry. In the other seminary, the dominion of prejudice is equally strong; and the works of Locke and Newton are the text from which the prelections are read. Mathematical learning is there the great object of study; but still we must disapprove of the method in which this subject is pursued.[18]

A few lines later, on the subject of mathematical learning at this second institution, easily recognisable as Cambridge, he adds, 'The invention finds no exercise; the student is confined within narrow limits; his curiosity is not aroused; the spirit of discovery is not awakened.'[18]

The charges made against the two universities were relatively fair ones. In the defence of Oxford, Edward Copleston, the Professor of Poetry, wrote an anonymous article, *A Reply to the Calumnies of the Edinburgh Review against Oxford* in 1810 and followed with *A Second Reply* (1810) and *A Third Reply* (1811). He identified the first half of the criticism as referring to his own university: 'The far-spread fame of Cambridge for mathematical studies fixes this description to Oxford.'[19]

Copleston had a fairly lengthy reply to all the points raised in the passage quoted from the *Edinburgh Review*, and for the purposes of this book, the most interesting is his reply to the charge that the student had no means of advancing beyond the mere elements of geometry.

What are the mere elements of Geometry? Are Plane and Spherical Trigonometry, are the properties of Conic Sections, of Conchoids, Cycloids, the Quadratrix, Spirals, &c. &c. the mere elements of Geometry? Is the method of Fluxions included under the same appellation? On all these subjects, lectures both public and private are given ...

And that the Scholar has the means of advancing to Newton's *Principia*, is tolerably proved by the public examination of Candidates for the Degree of B.A. twice every year, in that work.[20]

Copleston's two further 'Replies' and a second article by the writer in the *Edinburgh Review* in 1811, take up the question of just how elementary this mathematical tuition was. It may be said in conclusion that it is gratifying to know that Oxford was not unconcerned with some higher branches of mathematics at this time, but the content of the studies was certainly inferior to that of the contemporaneous course at Cambridge. No Oxford man at this time achieved any important mathematical status.

The situation at Cambridge was much more interesting. Mathematics was held in very high regard, and for the thirty years in question it was the dominant subject at undergraduate level. Towards the end of the eighteenth century, the Senate House examination had become the effective test for the Bachelor of Arts degree, and this took an increasingly mathematical form. It was stated in 1802 that 'For the future no degree should pass unless the candidate should have a competent knowledge of the first book of Euclid, arithmetic, vulgar and decimal fractions, simple and quadratic equations and Locke and Paley.'[21]

This supplied a minimum standard for the pass degree, but the honours candidates, who were selected in advance by their tutors, had a much harder examination than the one suggested by this quotation. It was regarded throughout the country as the supreme intellectual competition, and the top man, the Senior Wrangler, was held in great esteem.

Mathematics was so strongly entrenched at Cambridge that, even in 1824 when the first examinations in the classics tripos were held, it was stipulated that only those who had already obtained mathematical honours could take this second tripos. This rule remained until 1850. Samuel Butler lamented in 1822: 'I do not believe that in the days of Barrow, Newton, and Cotes the same exclusive attention was paid to mathematics as at the present time.'[22]

It might be thought that the great intellectual status placed on this course of study, and the content of the examination, would have been greatly beneficial to the progress of mathematics on a national scale. This proved not to be the case for a variety of reasons. The first was that the whole system of education and examination was one of self-perpetuation and any form of change, let alone research, received no official encouragement. Looking through the examination papers from 1801 to 1820 it is difficult to detect any major difference of content from beginning to end. Butler described the examination questions as 'A set of mathematical conundrums, in which each examiner tries to display his ingenuity by quibbling subtleties, by little niceties, and knackeries, and tricks of the art, which are for the most part exceedingly clever, and exceedingly unprofitable.'[23]

Charles Babbage described how in his first year at Cambridge he quickly discovered the futility of approaching a tutor for help with any problem which was outside the syllabus. He would only be told not to worry as such questions were never set in the examination.[24] William Whewell, describing the way that students prepared for this test, said:

> It is neither surprising nor blameable that our candidates for mathematical honours should be content with general formulae as the solutions of all problems, and general equations as the representatives of all principles, if they find that their Examiners appear to desire nothing more.[25]

Whewell made another remark in this book which unintentionally seems to point to another deficiency in the teaching method. Referring to mathematical publications written by former students he observed that a large proportion were concerned with the first principles of the differential calculus, and added:

> The frequency of such publications may, I think, be explained, by the consideration above stated; – that there prevails among us a mode of teaching the subject, in which the fundamental conception is slightly referred to or entirely eluded. For this being done, the impression left on the student's mind will necessarily be unsatisfactory; and when in a season of leisure, and with some skill in calculation, he begins to settle his own views, it is probable that some new aspect of the subject will appear to him more clear than that of his masters.[26]

One may equally well make the inference that the tutors did not themselves (with the exception of Woodhouse) sufficiently understand the first principles of the calculus to be able to teach it with confidence.

There was also the criticism that the mathematics taught at Cambridge served no useful purpose in contrast to the French, for example, who encouraged extensive mathematical study as a necessary preliminary for engineering applications, with great success both in theory and practice. Samuel Butler, in the tract already referred to, made a vigorous attack on the wastefulness of the whole Cambridge system, especially as it affected the pass degree men.

> On an average for the last three years, 146 men enter the senate-house annually, at the usual degree time.

Of these, 52 obtain honours; of whom 19 are wranglers, or proficients in mathematics; 19 are senior optimes or second-rate mathematicians; 14 are junior optimes; or smatterers.

What are the remaining 94? What have they to shew for an education of three years and a quarter, at an expense which cannot be short of £700? What have they got in religion, ethics, metaphysics, history, classics, jurisprudence? Who can tell? for, except the short examination of one day in Locke, Paley and Butler, in the senate-house, the University must be supposed to know nothing of their progress in these things. Their University examination for their degree is in mathematics, and if they have got four books of Euclid (or even less), can answer a sum in arithmetic, and solve a simple equation, they are deemed qualified for their degree, that is, the University pronounces this a sufficient progress, after three years and a quarter of study.[27]

Butler went on to make the claim that even after this exclusively mathematical course very few of the graduates ever pursued their studies in this subject. As far as the practical use of their knowledge was concerned he said: 'Take a junior or senior optime, or even a wrangler, into an irregular field with a common land-surveyor, and ask them severally to measure it: which will do it soonest and best?'[28]

I have already suggested that mathematics was at this time considered to be a useful mental training but to have little intrinsic importance. According to George Peacock,[29] the students at Cambridge could be divided into two classes: those with the wealth to follow no profession, and those intending to follow the professions of law, medicine or the church. The church was the most popular of these, at least half of the students intending to take Orders. There was very little possibility of a professional career in mathematics. The few positions as professor at any of the British universities or military academies were extremely difficult to obtain; even someone of the calibre of Charles Babbage got to hear of a possible vacancy only through well-informed friends and then, after a campaign in which he obtained testimonials from nearly all of the nation's leading mathematicians and had a manifesto printed, was still unsuccessful.[30]

John Toplis was quite blunt about this:

Perhaps one reason to be assigned for the deficiency of mathematicians and natural philosophers is the want of patronage. The sciences are so abstruse that, to excel in them, a student must give up his whole time, and that without any prospect of recompense; and should his talents and application enable him to compose a work of the highest merit, he must never expect, by publishing it, to clear one-half of the expense of printing.[31]

He proceeded to contrast this with the situation elsewhere:

In France and most other nations of Europe it is different: in them the student may look forward to a place in the National Institute or Academy of Sciences, where he will have an allowance sufficient to enable him to comfortably pursue his studies; and should he produce works worthy of publishing, they will be printed at the expense of the nation.[31]

William Whewell was, however, convinced of the indirect benefits to be obtained from a mathematical training.

No one who knows the recent history of this University can doubt that the mathematical studies of its members have produced a very powerful effect on the general character of their mental habits. Any one who is acquainted with the lawyers, or men of business, or statesmen, whom the University has produced in our own and the preceding generation, knows from observation of them and from their own declarations, that their mathematical pursuits here have in no small degree regulated their mode of dealing with other subjects. With respect to lawyers indeed, we have the evidence of their practical success to the reality of this connexion; and, among them, the extraordinary coincidence of professional eminence in after-life with mathematical distinction in their university career, shews that our studies may be an admirable discipline and preparation for pursuits extremely different from our own.[32]

There seems evidence then, that Cambridge-trained mathematicians could become good priests, doctors, or lawyers even if they could not measure fields. Whewell admits, however, that his views of the usefulness of mathematical training were by no means universally accepted:

On the one hand we are told that mathematics is a most admirable mental discipline; that it generates habits of strict reasoning, of continuous and severe attention, of constant reference to fundamental principles: on the other side it is asserted, that mathematical habits of thought unfit a man for the business of life; – make his mind captious, disputatious, over subtle, over rigid; that a person inured to mathematical reasoning alone, reasons ill on other subjects, seeks in them a kind and degree of proof which does not belong to them, becomes insensible to moral evidence, and loses those finer perceptions of fitness and beauty, in which propriety of action and delicacy of taste must have their origin.[33]

The usefulness, or otherwise, of mathematics as a mental training for professional life is a most interesting point of debate. It would appear that in the early nineteenth century very few thought of mathematics as anything but a mental discipline.

The Scottish universities had a much more liberal approach to education than the English ones at this time, but the standard of mathematical tuition was generally low. A Royal Commission on the Universities of Scotland in 1831 gave detailed information on this matter.[34]

The first reference is to the large classes in each university, where it was customary for one professor to do all the lecturing:

> In the Scotch Universities the Science of Mathematics is taught in public classes, sometimes numerously attended, where the individual students cannot expect to receive the personal attention necessary for fixing accurately in their minds the first elements.[35]

The subject was apparently taught in most Scottish schools:

> Few students enter the Mathematical class totally ignorant of the first principles of the science. Practical Mathematics are generally taught in all parts of the country, and the Parochial Schoolmasters are in general qualified to teach Mathematics, as far as the subject has hitherto been prosecuted, to the first classes of the Universities.[35]

Edinburgh University had three mathematical classes with an average altogether of 188 students per year for the six years preceding 1826. In the year 1825/26, there were seventy-three students in the first class, fifty-seven in the second and twelve in the third. The average age in the first class was between fourteen and fifteen years.

> It meets for an hour each day during the session of five months and a half. In this class the Elements of Geometry are taught, and afterwards Plane Trigonometry, with its application to the Mensuration of heights and distances, also Mensuration in general. The second class, which some students enter without attending the first, and which many who did attend the first do not join, study Algebra, with its various applications, the Cal-

culus of Signs, Conic Sections, Solid Geometry, Spherical Trigonometry, Dialling, and as many more branches as can be overtaken in the course of the Session.[36]

In view of the age of the students in these two classes, it is hardly surprising that the content of the course is relatively small. The third class, however, which was intended mainly for older students, failed to provide the stimulus which the first two also lacked.

> In the third class, which is frequently attended by persons of a different description from regular Students, such as practising Engineers, and sometimes Officers of the Army, and which, therefore, is not so definite in its constitution as to age as the other two, some of the subjects taught in the second class are resumed; but the main object is the Differential Calculus, or Doctrine of Fluxions, with its applications.[36]

The situation at Glasgow University was fairly similar. With reference to the Professor of Mathematics:

> In the junior class he teaches the Elements of Arithmetic and Algebra, the first six books of Euclid, with Mensuration, Plane Trigonometry and Surveying. In the second class he teaches Conic Sections, the 11th and 12th books of Euclid, Spherical Trigonometry and Navigation, also the higher rules of Algebra, with the Elements of Fluxions.[37]

The two classes each received one hour's tuition per day for five days a week.

The course at King's College, Aberdeen, consisted of arithmetic, algebra, geometry, plane trigonometry and mensuration.[38] Marischal College, Aberdeen, had a more advanced course which consisted of these subjects, together with cubic and biquadratic equations, solid geometry, navigation and spherical trigonometry. In the third year, higher equations, applications of algebra to geometry, indeterminate problems, natural philosophy and fluxions were studied.[39]

The mathematics at St Andrews compared well with that at Marischal,[40] but the Commission recommended that at both of these institutions the elementary mathematics taught in the first year could profitably be omitted and taught instead at school level.

We may conclude that mathematics teaching was by no means neglected at universities during this period, especially at

Cambridge and in Scotland, but there is no indication of these places being centres of research, and little to suggest that they were in a position to stimulate progress which compared with that being made on the Continent.

The next section of this chapter can be suitably prefaced with another quotation from the writer in the *Edinburgh Review*:

> Perhaps, too, we might allege, that another public institution, intended for the advancement of science, the Royal Society, has not held out, in the course of the greater part of the last century, sufficient encouragement for mathematical learning.[41]

During the early part of the nineteenth century, the Royal Society was subjected to a mounting criticism which culminated in 1830 in the publication of two books, Charles Babbage's *Observations on the Decline of Science in England* and A. B. Granville's anonymously written *Science without a Head*. These two books, while disagreeing on many points, both strongly supported the contention that the Royal Society was largely to blame for much that was wrong in British science. The major charge was that the genuine scientists found themselves in a small minority in a Society which had become composed largely of an assortment of bishops, noblemen, service officers, clergy and lawyers. It was too easy to become a member of this body which was supposedly devoted to the advance of science in Britain. All that was required was for three Fellows to sign a certificate vouching for the usefulness of the candidate to the Society, and, if no opposition came for a period of ten weeks, he then became a Fellow. There was no limit to the number of Fellows at any time. This state of affairs continued until the statutes were revised in 1847, when stricter rules were observed in the election of Fellows.[42] Granville analysed the membership for the period 1800–30 and found that out of a total of 641 Fellows, only 103 had made scientific contributions in the form of a paper for the Society's journal, *Philosophical Transactions*.[43] Amongst these, the ten bishops in the Society had produced nine papers all by J. Brinkley, the Lord Bishop of Cloyne, and the sixty-three lords no papers at all. In an attempt to establish government of the Royal Society by scientists of great reputation, J. F. W. Herschel was nominated for President in 1830 in opposition to the Duke of Sussex but was defeated by 119 votes to 111.[42]

The *Philosophical Transactions* was one of the small number of scientific journals appearing in Britain at this time. The

Philosophical Magazine, T. Thomson's *Annals of Philosophy* and the *Journal of Science and the Arts* produced by the Royal Institution of Great Britain devoted themselves to science and technology with very few mathematical articles ever appearing. The *Edinburgh Philosophical Journal* began in 1819 and the *Transactions of the Cambridge Philosophical Society* in 1821. Whether or not the Royal Society had done much to encourage mathematical learning, the mathematicians had certainly little to offer the *Philosophical Transactions.* Out of a total of 760 papers published in this journal between 1800 and 1830, only forty-four were mathematical, the output of twenty mathematicians. From these forty-four, one was written by a Frenchman, three were concerned with problems of life insurance, the contents of two more were considerably out of date even for that period and at least four others were of a relatively elementary nature. It may be concluded that only about thirty-five (an average of just over one per year) of these papers had genuine original content. Amongst these, the most notable contributors were Thomas Knight with seven papers, Robert Woodhouse, James Ivory and Charles Babbage with four each, J. F. W. Herschel with three and a single paper by W. G. Horner. The most popular subjects were geometry (8), mechanics (5) and infinite series (9). The geometrical papers were mostly classical in content indicating that British mathematicians were at this time as unfamiliar with Descartes as they were with Leibniz. The ones on infinite series were most ambitious with attempts to sum the most incredible series, but with little or no technique for dealing with questions of convergence. It is true that little was known about work in this field on the Continent, but even the discoveries of the eighteenth-century Englishman E. Waring seem to have been largely ignored. The papers on mechanics, all but one by James Ivory, were of a higher standard and developed Laplace's work on potential, considering the attractions of spheroids and ellipsoids.

Amongst the other papers, two by Woodhouse and concerned with fundamental principles are of interest. In one of these he analysed the difference between algebraic and geometrical methods of solving a problem, concluding that the latter may have had a greater aesthetic appeal but the former was much more useful. In the other he considered the logic of the use of imaginary quantities, asserting the falsity of argument, by analogy in a mathematical context.

Two papers by Charles Babbage 'Essay towards the calculus of functions', Parts I and II, written in 1815 and 1816, were both much longer than the normal paper. These are discussed in chapter 4. The writer ambitiously tried to establish a new branch of mathematics in a systematic way, showing how to solve some very complicated functional equations.

Finally, the only paper in this collection that has survived the test of time, 'New methods of solving numerical equations of all orders', written in 1819 by W. G. Horner, gave in a somewhat involved way the 'Horner's method' used in numerical analysis.

As far as books are concerned, it can be said with some confidence that not more than forty mathematical works were written in Britain during these thirty years and that hardly any of these made an original contribution. Textbooks on geometry were very popular and, in nearly all of these, analytical methods were ignored. The only new thing that came out of any of these, and which still survives, is Playfair's axiom: 'Two straight lines which intersect one another, cannot be both parallel to the same straight line.'[44] It having already been defined that 'Parallel straight lines are such as are in the same plane, and which, being produced ever so far both ways, do not meet.'[45]

P. Barlow wrote on the Theory of Numbers[46] in 1811 in which he solved some indeterminate equations of up to the third and higher degrees, and introduced to British readers the work of Gauss and Legendre. He also compiled a mathematical dictionary in 1814.[47] Other works included textbooks on conics, logarithms and trigonometry by D. M. Peacock, on the solution of algebraic equations by J. Ivory and W. Spence, on mechanics by J. Playfair and W. Whewell, two volumes on the geometrical properties of complex numbers by B. Gompertz and two books on mathematical applications to various facets of life insurance by F. Baily. Undoubtedly the best writer at this time was Woodhouse, who made contributions to the foundations of calculus,[48] trigonometry,[49] and the calculus of variations.[50] Of these three, the book on trigonometry was mostly a textbook and the one on the calculus of variations a historical study of isoperimetrical problems. The first book had a quality far above that of his contemporaries and it is a pity that it did not have a more immediate influence on British mathematics.

The thirty-page preface made four important points. The first was an attack on the idea of fluxions which throughout the previous century had been basic to British concepts of the calculus. Woodhouse insisted that an analytical process such as the differential calculus should be explicable by analytical principles only, and that it was totally wrong to base it instead on intuitive ideas of motion, as Maclaurin and even Newton appeared to have done.

Secondly, he took up Berkeley's contention about the 'shifted hypothesis' and applied it to various attempts by limits to establish the calculus. He showed that all arguments depending on this ill-defined principle were equivalent to stating that a result was true provided a certain variable quantity was non-zero at one stage of the argument and then equal to zero later on. Such inconsistencies as Berkeley had already shown could not be tolerated in an argument that claimed to be logical.

Thirdly, he analysed Lagrange's formal attempt to define the successive derivatives of a function from the coefficients of the Taylor expansion of that function with close scrutiny and rejected his argument. Lagrange's idea[51] had been to avoid all appeal to limits, fluxions, vanishing quantities or infinitesimals by the expedient of giving an algebraic proof of the general formula for the Taylor series and then defining his derivatives accordingly. Woodhouse, with some quite elegant analysis, succeeded in demonstrating that Lagrange had employed circular arguments, used unjustified assumptions and brought in unintentionally the concept of the limit which he had been studiously trying to avoid. I would add that Lagrange also used intuitive ideas of continuity and even 'shifted his hypotheses'. Woodhouse's purpose in demolishing Lagrange's argument was to clear the ground for an attempt by himself to establish the calculus on a similar basis. The rest of the book is devoted to such an attempt. He tried first to prove the binomial theorem algebraically, and expanded other functions by means of the binomial series before arriving at a fairly general Taylor series. I can only value his analysis as a brave try. There is much to be commended in his effort, but, even allowing for the inadequate knowledge of convergence essential to his argument, he still seemed in several places to have fallen below his own standards of logic and to have made many of the errors he so clearly exposed in the work of others.

Having satisfied himself that he had arrived at a proof of Taylor's expansion, his subsequent application was excellent. He showed how all the normal rules for manipulating the calculus could be demonstrated; he defined velocity and the gradient of a tangent to a curve rigorously by means of first derivatives, thus completing his criticism of the fluxionists who tried to put the cart before the horse; he gave an accurate definition of a differential in terms of a derivative; and he went near to giving an analytical definition of continuity. At any rate he was probably the first to show that continuity must be defined in terms of limits and not to define a limit by appealing to a vague intuitive 'law of continuity'.

His fourth idea was to use the differential notation of Leibniz throughout the book, in preference to the fluxionary symbols of Newton. This was remarkable as being probably the first example of a British mathematical work for over a century to depart from the traditional notation and change to the more useful and suggestive type used on the Continent. It is described in more detail in the next chapter.

A totally different source of mathematical productivity at this time was a series of bulky volumes edited by F. Maseres, the Cursitor Baron of the Court of Exchequer.[52] The tracts contained in these volumes dealt with the solution of equations by exact and approximate methods, logarithms, and infinite series, with applications to navigation and insurance. It is difficult to assess the originality of these works in relation to the other material under consideration since most of the papers were undated and many were reprints of older works by mathematicians like Halley, Pascal, Simson, Newton, Cardan, Descartes and Harriot.

From the evidence that has been presented it is clear that there was a weakness, a lack of motivation and a misunderstanding of the nature and possibilities of the subject at all levels of British mathematics in the early nineteenth century. The major instrument for reform at this time was the Analytical Society, and as Babbage played a very prominent part in this group I now consider their activities in detail, especially the introduction of the differential notation to Great Britain.

NOTES

1. A. B. Granville, *Science without a Head*, London, 1830, p. 20.
2. *Edinburgh Review*, 1807, **11**, 249–84.
3. *Ibid.*, p. 281.
4. Rev. J. Toplis, *Philosophical Magazine*, 1805, **20**, 25.

5. Rev. Baden Powell, *An Historical View of the Progress of the Physical and Mathematical Sciences from the Earliest Ages to the Present Times*, London, 1834, p. 363.
6. R. Woodhouse, *Principles of Analytical Calculation*, Cambridge, 1803, p. xxvii.
7. Woodhouse, *op. cit.*, p. xvii.
8. W. Ludlam, *The Rudiments of Mathematics*, Cambridge, 1785, p. 5.
9. *Edinburgh Review, op. cit.*, p. 282.
10. Baden Powell, *op. cit.*, pp. 367–8.
11. *Report of Her Majesty's Commissioners appointed to inquire into The Revenues and Management of certain Colleges and Schools, and the Studies pursued and Instruction given therein*, London, 1864.
12. *Ibid.*, p. 81.
13. *Ibid.*, pp. 214–15.
14. *Ibid.*, p. 246.
15. *Ibid.*, p. 252.
16. *Report of the Commissioners appointed to inquire into the State of Popular Education in England.* (*Reports, Commissioners*, 1861, **21**, vol. I, pp. 660–9.)
17. F. Walkingame, *The Tutor's Assistant*, London, 1751, p. 147.
18. *Edinburgh Review, op. cit.*, p. 283.
19. E. Copleston, *A Reply to the Calumnies of the Edinburgh Review against Oxford*, Oxford, 1810, p. 15.
20. *Ibid.*, pp. 18–19.
21. *Cambridge University Calendar*, 1802, p. xxxviii.
22. Eubulus (Rev. S. Butler), *Thoughts on the Present System of Academic Education in the University of Cambridge*, London, 1822, p. 11.
23. *Ibid.*, p. 9.
24. C. Babbage, *Passages from the Life of a Philosopher*, London, 1864, pp. 26–7.
25. W. Whewell, *Thoughts on the Study of Mathematics as Part of a Liberal Education*, Cambridge and London, 1836, p. 43.
26. *Ibid.*, p. 41.
27. Butler, *op. cit.*, pp. 5–6.
28. *Ibid.*, p. 8.
29. G. Peacock, *Observations on the Statutes of the University of Cambridge*, London, 1841.
30. C. Babbage, British Museum Additional Manuscripts 37182, No. 54.
31. Toplis, *op. cit.*, p. 27.
32. Whewell, *op. cit.*, p. 40.
33. *Ibid.*, p. 3.
34. *Royal Commission of Inquiry into The State of the Universities of Scotland*, 1831. (Reports, Commissioners, 1831, 12.)
35. *Ibid.*, p. 31.
36. *Ibid.*, pp. 126–7.
37. *Ibid.*, p. 246.
38. *Ibid.*, p. 320.
39. *Ibid.*, pp. 351–2.
40. *Ibid.*, pp. 401–2.

41. *Edinburgh Review, op. cit.*, p. 283.
42. E. N. da C. Andrade, *A Brief History of the Royal Society*, London, 1960.
43. Granville, *op. cit.*, p. 34.
44. J. Playfair, *Elements of Geometry containing the first six books of Euclid*, Edinburgh, 1814, p. 23.
45. *Ibid.*, p. 22.
46. P. Barlow, *An Elementary Investigation of the Theory of Numbers*, London, 1811.
47. P. Barlow, *A New Mathematical and Philosophical Dictionary*, London, 1814.
48. R. Woodhouse, *Principles of Analytical Calculation*, Cambridge, 1803.
49. R. Woodhouse, *Trigonometry*, Cambridge, 1809.
50. R. Woodhouse, *Isoperimetrical Problems*, Cambridge, 1810.
51. J. L. Lagrange, *Théorie des Fonctions Analytiques*, Paris, 1797.
52. F. Maseres, *Scriptores Logarithmici*, London, 1801, 1804, 1807, vols. 4, 5, 6.

3

The Analytical Society

As mentioned in the previous chapter, probably the first British book for over a century to make consistent use of the differential notation was Robert Woodhouse's *Principles of Analytical Calculation*, Cambridge, 1803. A Scottish mathematician, James Ivory, also used this symbolism in his paper 'On the attractions of homogeneous ellipsoids', *Philosophical Transactions*, 1809, **99**, 345–72.

It is doubtful if either Woodhouse or Ivory was able to influence their colleagues to adopt the Continental notation. This was achieved by means of a vigorous campaign a few years later by younger men, most notably J. F. W. Herschel, George Peacock and Charles Babbage. They helped to form the Analytical Society, whose objective was to reform British mathematics generally, starting with notation.

The best account of the early days of this movement is given in Babbage's autobiography *Passages from the Life of a Philosopher*, London, 1864. In an early chapter he describes his undergraduate days at Cambridge which he entered in 1811, eight years after the publication of Woodhouse's book. He very quickly became disillusioned with the mathematics teaching at the university, finding it to be of lower standard than his own private study. He says: 'Thus it happened that when I went to Cambridge I could work out such questions as the very moderate amount of mathematics which I then possessed admitted, with equal facility, in the dots of Newton, the d's of Leibnitz, or the dashes of Lagrange.'[1] This familiarity with notations was evidently not taught at the university, despite Woodhouse's book, and, more surprisingly, his presence as a teacher. Babbage continues:

31

> I thus acquired a distaste for the routine of the studies of the
> place, and devoured the papers of Euler and other mathemati-
> cians scattered through innumerable volumes of the academies
> of Petersburgh, Berlin, and Paris, which the libraries I had
> recourse to contained.
>
> Under these circumstances it was not surprising that I should
> perceive and be penetrated with the superior power of the
> notation of Leibnitz.[2]

He goes on to describe how he read Woodhouse's *Principles of
Analytical Calculation*, which helped to teach him the Leibnizian
notation. This is the only reference made to Woodhouse in this
movement, and the indication is that Woodhouse himself took no
part in this purely undergraduate activity. Babbage also read
Lagrange's *Théorie des Fonctions Analytiques* and described how he
wanted above all to obtain a copy of the standard French textbook
on the differential and integral calculus by Lacroix, which he
eventually procured despite the difficulties caused by the war
(which made all French books very scarce) and the almost pro-
hibitive price of seven guineas. This book so impressed him that
he wanted to have it translated into English.

> I then drew up the sketch of a society to be instituted for
> translating the small work of Lacroix on the Differential and
> Integral Lacroix [Calculus]. It proposed that we should have
> periodical meetings for the propagation of d's; and consigned
> to perdition all who supported the heresy of dots. It maintained
> that the work of Lacroix was so perfect that any comment was
> unnecessary.[2]

As a result of discussions with two of his friends, Slegg and
Bromhead, he decided to form with them a society for the
cultivation of mathematics. This was the group which became the
famous Analytical Society and we shall see from the next quotation that there were at least nine members at first, but for one
reason and another the effective work was done by the three most
eminent members, Herschel, Babbage and later Peacock.

> At that meeting, besides the projectors, there were present
> Herschel, Peacock, D'Arblay, Ryan, Robinson, Frederick Maule
> and several others. We constituted ourselves 'The Analytical
> Society'; hired a meeting-room, open daily; held meetings, read
> papers and discussed them. Of course we were much ridiculed
> by the Dons; and, not being put down, it was darkly hinted that
> we were young infidels, and that no good would come of us.

In the meantime we quietly pursued our course, and at last resolved to publish a volume of our Transactions. Owing to the illness of one of the number, and to various other circumstances, the volume which was published was entirely contributed by Herschel and myself.

At last our work was printed, and it became necessary to decide upon a title. Recalling the slight imputation which had been made upon our faith, I suggested that the most appropriate title would be –

'The Principles of pure D-ism in opposition to the Dot-age of the University.'[3]

The Analytical Society produced three publications. The first, called *Memoirs of the Analytical Society*, was produced in 1813. This was followed by a translation of Lacroix's *Sur le Calcul Différentiel et Intégral* in 1816 and a book of examples on calculus in 1820.

The first work, according to Babbage, was written entirely by Herschel and himself. The papers on topics such as continued products and difference equations are of a very high standard for those times, but for our purposes the most interesting part is the preface. This is a general essay on mathematics giving a historical survey up to the time of writing. It is remarkable that the works of no fewer than thirty-five distinguished mathematicians are quoted and commented upon with authority, a fact that indicates the unusual depth of the reading of these two undergraduates and their friends.

In view of the aims of the society it is interesting to see what they have to say about symbolism, and an elegant description of the importance of good notation in mathematics is given at the beginning:

Three causes however chiefly appear to have given so vast a superiority to Analysis, as an instrument of reason. Of these, the accurate simplicity of its language claims the first place. An arbitrary symbol can neither convey, nor excite any idea foreign to its original definition. The immutability, no less than the symmetry of its notation, (which should ever be guarded with a jealousy commensurate to its vital importance,) facilitates the translation of an expression into common language at any stage of an operation, – disburdens the memory of all the load of the previous steps, – and at the same time, affords it a considerable assistance in retaining the results. Another, and perhaps not less considerable cause, is to be found in the conciseness of that notation.[4]

Again – this point is emphasised with a sharp criticism of existing notation: 'To employ as many symbols of operation and as few of quantity as possible, is a precept which is found invariably to ensure elegance and brevity. The very reverse of this principle forms the character of symbolic analysis, up to within fifty years of the present data.'[5]

The authors go on to discuss the history of calculus and comment on the shabbiness of the eighteenth-century controversies. 'It is a lamentable consideration, that that discovery which has most of any done honour to the genius of man, should nevertheless bring with it a train of reflections so little to the credit of his heart.'[6]

On the same page Herschel and Babbage describe the calculus as 'Discovered by Fermat, concinnated and rendered analytical by Newton, and enriched by Leibnitz with a powerful and comprehensive notation' and 'as if the soil of this country were unfavourable to its cultivation, it soon drooped and almost faded into neglect, and we have now to re-import the exotic, with nearly a century of foreign improvement, and to render it once more indigenous among us'. Here we have a rarity among English mathematicians of this time, a treatise on calculus devoid of all partisan prejudice. It is very bold for these writers actually to attribute the invention to Fermat and put the achievements of Newton and Leibniz into a more proper perspective. This is a great improvement on the barren Newton *v.* Leibniz controversy, and the authors rightly point out that this diversion away from developing Newton's calculus resulted in a century of regress for British mathematics.

Their last word on notation before presenting their treatises is:

> The importance of adopting a clear and comprehensive notation did not, in the early period of analytical science, meet with sufficient attention; nor were the advantages resulting from it duly appreciated. In proportion as science advanced, and calculations became more complex, the evil corrected itself, and each improvement in one, produced a corresponding change in the other. Perhaps no single instance of the improvement or extension of notation better illustrates this opinion, than the happy idea of defining the result of every operation, that can be performed on quantity, by the general term of function, and expressing this generalization by a characteristic letter. It had the effect of introducing into investigations, two qualities once deemed incompatible, generality and simplicity.[7]

Peacock assisted Herschel and Babbage in their production of the other two publications. Babbage describes their motives and the way in which the labour was divided between them, in his autobiography. It appears that they chose to translate a French work and compose a set of examples as the best method to enforce their views on the university. What Babbage says, here, again reflects poorly on the mathematical teaching offered at Cambridge.

> My views respecting the notation of Leibnitz now (1812) received confirmation from an extensive course of reading. I became convinced that the notation of fluxions must ultimately prove a strong impediment to the progress of English science. But I knew, also, that it was hopeless for any young and unknown author to attempt to introduce the notation of Leibnitz into an elementary work. This opinion naturally suggested to me the idea of translating the smaller work of Lacroix. It is possible, although I have no recollection of it, that the same idea may have occurred to several of my colleagues of the Analytical Society, but most of them were so occupied, first with their degree, and then with their examination for fellowships, that no steps were at that time taken by any of them on that subject.
>
> Unencumbered by these distractions, I commenced the task, but at what period of time I do not exactly recollect. I had finished a portion of the translation, and laid it aside, when, some years afterwards, Peacock called on me in Devonshire Street, and stated that both Herschel and himself were convinced that the change from the dots to the d's would not be accomplished until some foreign work of eminence should be translated into English. Peacock then proposed that I should either finish the translation which I had commenced, or that Herschel and himself should complete the remainder of my translation. I suggested that we should toss up which alternative to take. It was determined by lot that we should make a joint translation. Some months after, the translation of the small work of Lacroix was published.
>
> For several years after, the progress of the notation of Leibnitz at Cambridge was slow. It is true that the tutors of the two largest colleges had adopted it, but it was taught at none of the other colleges.
>
> It is always difficult to think and reason in a new language, and this difficulty discouraged all but men of energetic minds. I saw, however, that, by making it their interest to do so, the change might be accomplished. I therefore proposed to make a

large collection of examples of the differential and integral calculus, consisting merely of the statement of each problem and its final solution. I foresaw that if such a publication existed, all those tutors who did not approve of the change of the Newtonian notation would yet, in order to save their own time and trouble, go to this collection of examples to find problems to set to their pupils. After a short time the use of the new signs would become familiar, and I anticipated their general adoption at Cambridge as a matter of course.

I commenced by copying out a large portion of the work of Hirsch. I then communicated to Peacock and Herschel my view, and proposed that they should each contribute a portion.

Peacock considerably modified my plan by giving the process of solution to a large number of the questions. Herschel prepared the questions in finite differences, and I supplied the examples to the calculus of functions. In a very few years the change was completely established; and thus at last the English cultivators of mathematical science, untrammelled by a limited and imperfect system of signs, entered on equal terms into competition with their continental rivals.[8]

The campaign was carried a stage further by Peacock. He was appointed moderator for the examinations in mathematics at Cambridge on three occasions, and determined to use this opportunity to help his cause by setting his questions on calculus in the differential rather than the prevailing fluxionary notation.

Luckily, we possess copies of the questions set in these examinations, and this collection is one of the earliest sets of such examination problems in existence. Published by Black and Armstrong, the foreign booksellers to the king, the title of the book is: *Cambridge problems: being a collection of the printed questions proposed to the Candidates for the Degree of Bachelor of Arts at the General Examinations, from the year* 1801 *to the year* 1820 *inclusively*, London, 1836. It is probable that previous to this time the questions were dictated orally at the examinations. According to W. W. R. Ball in *A History of the Study of Mathematics at Cambridge*,

> About this time, circ. 1772, it began to be the custom to dictate some or all of the questions and to require answers to be written. Only one question was dictated at a time, and a fresh one was not given out until some student had solved that previously read – a custom which by causing perpetual interruptions to take down new questions must have proved very harassing.[9]

For this reason no printed copies of examinations before 1801 exist and it is highly fortunate that the period of the first recorded examinations, 1801–20, is precisely the one which is most relevant to this chapter.

The questions cover an immense range of topics, and, while some are trivially simple, others are very difficult. The topics examined include Euclidean geometry, arithmetic and elementary algebra, plane and spherical trigonometry, analytical geometry of two and three dimensions, conic sections, differential and integral calculus, differential equations, finite differences and series, the calculus of variations, statics, dynamics, astronomy, central forces, hydrostatics, optics and the theory of attractions. Applied mathematics is preferred to pure in the ratio of roughly four questions to three, and in the pure mathematics a disproportionate amount is given to arithmetical problems, while strangely, in view of the English predilection for it as a 'discipline' subject, Euclidean geometry is almost neglected. There are questions like finding what part of half a crown is equal to $\frac{3}{7}$ of 1s. $5\frac{1}{2}$d. or enumerating the different practical methods of determining the latitude of a ship at sea, but the calculus problems, as will be shown, are quite formidable.

Throughout this period two moderators were appointed each year, responsible for setting, invigilating and marking the examinations, and the book of papers gives their names. It is interesting to observe that Woodhouse was one of the moderators for the years 1803, 1804, 1807 and 1808. The fact that he was chosen for these years indicates that the university bore no grudge against him for the indiscretions of his *Principles of Analytical Calculation*, published in 1803. It also shows that Woodhouse did not carry his principles about notation into his examination responsibility, for all the calculus questions set in these years are in the fluxionary notation.

It is also interesting to note that the office of moderator was not normally given to a senior member of the university, but entrusted to more youthful lecturers. Of the only three among this list of moderators who achieved any subsequent degree of fame, Woodhouse first became moderator at the age of 29, Peacock at 25 and William Whewell at 25. It appears, however, that moderating was a job requiring great physical as well as mental stamina. In a letter to his sister dated 16 January 1820, Whewell writes:

The examination in the Senate House begins to-morrow, and is rather close work while it lasts. We are employed from seven in the morning till five in the evening in giving out questions and receiving written answers to them; and when that is over, we have to read over all the papers which we have received in the course of the day, to determine who have done best, which is a business that in numerous years has often kept the examiners up the half of every night; but this year is not particularly numerous. In addition to all this, the examination is conducted in a building which happens to be a very beautiful one, with a marble floor and a highly ornamented ceiling; and as it is on the model of a Grecian temple, and as temples had no chimneys, and as a stove or a fire of any kind might disfigure the building, we are obliged to take the weather as it happens to be, and when it is cold we have the full benefit of it – which is likely to be the case this year. However, it is only a few days and we have done with it.[10]

The Spartan qualities required by this office, and the displeasure which certain holders gave to the authorities in the university, suggest the possibility of arguing a case that the post of moderator was intended as an academic punishment rather than a privilege! Peacock, for example, aroused much anger and controversy by the probems he set when first appointed moderator in 1817, and yet was given the job again in 1819 and 1821.

Until Peacock was moderator, all the questions on calculus had been set in the fluxionary notation. In 1816, the year before Peacock's office, the calculus questions were:[11]

Required fluents of

$$\frac{mbx^{-n-1}\dot{x}}{(e+fx^{-n})}, \quad \frac{\sqrt{(y)}\dot{y}}{1+y^{3/2}}, \quad \frac{\dot{x}}{\sqrt{(x^2-a^2)}},$$

$$\frac{x^{1/2}\dot{x}}{\sqrt{(2a-x)}}, \quad \frac{z\dot{z}}{\sqrt{(a^4-z^4)}}.$$

Find fluents of

$$\frac{\mathrm{d}\dot{x}}{a^3-mx^2}, \quad \frac{\mathrm{d}\dot{x}\sqrt{(x)}}{\sqrt{(1-x)}}, \quad \frac{x^2\dot{x}}{(x-a)^2(x+a)}.$$

Find fluents of

$$\frac{(a+bx)\,\mathrm{d}\dot{x}}{x^3-1}, \quad \frac{\dot{x}}{x^n\sqrt{(a^2-x^2)}}$$

when n even and

$$\frac{d\dot{x}}{x(1+x)^2(1+x+x^3)}.$$

Find fluent of

$$\frac{\dot{z}}{z\sqrt{(a+cz^n)}}; \quad \frac{x^p\dot{x}}{(1+x^n)^2},$$

the fluent of

$$\frac{x^p\dot{x}}{1+x^n}$$

being A.

If $ay^m\dot{y} = cx^n\dot{y} - ayx^{n-1}\dot{x}$, determine the algebraic relation of x and y.

Find fluents of

$$\frac{(a+bx)\dot{x}}{a^2+x^2} \quad \text{and} \quad \frac{x^4\dot{x}}{\sqrt{(a^2+x^2)}}.$$

Given

$$\sqrt{\frac{(x+a)^3}{x-a}} = \text{a minimum}.$$

Find value of x.

Required fluxion of arc whose sine $= 2y\sqrt{(1-y^2)}$.

Find fluents

$$\int \frac{x^4\dot{x}}{\sqrt{(a^2-x^2)}}, \int \frac{2a\dot{x}}{x\sqrt{(a^3+x^3)}}; \quad \int \theta\cos^3\theta.$$

Find fluents

$$\int \frac{x\dot{x}}{(1+x)^3(1+x+x^2)^{1/2}}, \int bx^{3/2}\dot{x}\frac{\sqrt{(2a-x)}}{(a-x)^2},$$

$$\int az^2y^{n-1}\dot{y} \quad \text{if } z = (b+cy^n)^m\dot{y}.$$

Solve fluxional equations $(x-y)\dot{x}\dot{y} = y\dot{x}^2 + (a-x)\dot{y}^2$ and

$$\frac{\ddot{x}}{\dot{z}^2} + x + \cos mz = 0.$$

There is a curious mixture of notation in these examples with Leibnizian symbols appearing here and there, but the general presentation of this notation is most unsystematic. It should also be remarked that at least half of these twenty-six problems are extremely complicated and difficult.

The following year, 1817, George Peacock was first appointed moderator, together with John White of Caius. Both followed the custom of setting papers independently of the other and, while White followed the traditional notation, Peacock's questions show great differences. Here are some examples of White's problems:[12]

Find fluents of

$$\frac{p\dot{x}}{\sqrt{(x)}\sqrt{(a-bx)}} \quad \text{and} \quad \frac{\dot{x}}{x\sqrt{(1+\sqrt{x})}}.$$

Find fluent of

$$\frac{\dot{x}}{(a^2+x^2)^{\frac{2p+1}{2}}}$$

and in fluent of

$$\frac{xm\dot{x}}{x^n - px^{(n-1)} + qx^{(n-2)} - \text{etc.}}$$

m being greater than n, show that the coefficients of all terms involving higher powers of x than the $(m-n+1)$th will vanish. Find fluents of

$$\frac{p\dot{x}}{a-bx^2} \quad \text{and} \quad \frac{x^3\dot{x}}{\sqrt{(a-x)}}.$$

Peacock's questions, however, included[12]

Find integral or fluent of

$$\frac{dx}{x^3 - 7x^2 + 12x}.$$

Integrate the differentials

$$\frac{dx}{x^5\sqrt{\{1+x^2\}}}, \quad \frac{a^x\,dx}{x^2} \quad \text{and} \quad dx\cos^2 x \sin^3 x.$$

Integrate the differences or increments x^3 and $e^x \cos x\theta$
Integrate the different equations

(1) $\dfrac{d^2 y}{dx^2} = \dfrac{m}{(a-y)^2}$,

(2) $d^2 y + Ay\,dx^2 = X\,dx^2$ where X is a function of x,

(3) $\dfrac{dz}{dx} - \dfrac{y}{x}\dfrac{dz}{dy} = -\dfrac{y^2}{x^2}$.

It is only in the differential equation problems that the differential notation is clearly asserted, but the other questions show a consistent use of Leibnizian symbols in contrast to the slapdash formulations of the previous year.

Peacock's accomplishment did not meet with general approval from the university. William Whewell, writing to John Herschel on 6 March 1817, mentions the controversy aroused:

> You have I suppose seen Peacock's examination papers. They have made a considerable outcry here and I have not much hope that he will be moderator again. I do not think he took precisely the right way to introduce the true faith. He has stripped his analysis of its applications and turned it naked among them. Of course all the prudery of the university is up and shocked at the indecency of the spectacle. The cry is 'not enough philosophy'. Now the way to prevent such a clamour would have been to have given good, intelligible but difficult physical problems, things which people would see they could not do their own way, and which would excite curiosity sufficiently to make them thank you for your way of doing them. Till some one arises to do this, or something like it, they will not believe, even though one were translated to them from the French.[13]

The last sentence quoted suggests a lack of immediate success on the publication of the translation of Lacroix's work by Peacock, Babbage and Herschel.

Peacock's own attitude is given in a letter he wrote to an unknown friend quoted in his obituary notice of the *Proceedings of the Royal Society* in 1859:

> I assure you, my dear——, that I shall never cease to exert myself to the utmost in the cause of reform, and that I will never decline any office which may increase my power to effect it. I am nearly certain of being nominated to the office of Moderator in the year 1818–19, and as I am an examiner in virtue of my office, for the next year I shall pursue a course even more decided than hitherto, since I shall feel that men have been prepared for the change, and will then be enabled to have acquired a better system by the publication of improved elementary books. I have considerable influence as a lecturer and I will not neglect it. It is by silent perseverance only that we can hope to reduce the many-headed monster of prejudice, and make the University answer her character as the loving mother of good learning and science.[14]

Peacock was in fact nominated as moderator for the 1819 examination, but before this Fearon Fallows and William French moderated in 1818. They ignored Peacock's reform and set all the calculus questions in fluxionary symbolism. Their only concession was to use the ∫ symbol for integration, or rather for finding fluents, all the d's being firmly suppressed and dots used liberally. To give some examples from their papers:[15]

To find the fluent of

$$\frac{\dot{x}\sqrt{(1-x^2)}}{(1+x)^2}.$$

Required fluxion of

$$\frac{y\sqrt{(a^2-y^2)}}{a^3-y^3}.$$

Find fluents

$$\int x^2\dot{x}\ \text{arc(sin}=x),\ \int\frac{\dot{x}(1+x^2)}{(1-x^2)\sqrt{(1+x^4)}}.$$

Find the relation of x and y in the following equations:

$$x\dot{y}-y\dot{x}-(x^2+1)\dot{x}=0;\ (\dot{x}^2+\dot{y}^2)^{3/2}+2a^{1/2}\sqrt{(a-x)}\dot{x}\ddot{y}=0.$$

For the 1819 moderation, Peacock was joined by Richard Gwatkin who had graduated from St John's as recently as 1814 and was more favourable to the new notation. Whewell writes in a letter to Herschel, dated November 1818: 'our moderators are Gwatkin and Peacock; in G. I have great confidence, he is cautious but reasonable, and he has been reading a good deal of good mathematics.'[16]

The cautious side in Gwatkin's nature was evident in the questions he set for 1819. While Peacock set, characteristically, problems like[17]

Integrate

$$(1)\ \frac{dx}{x\sqrt{(bx^2-a)}}\ \text{and}\ \frac{d\theta}{(\sin\theta)^4\cos\theta},$$

$$(2)\ \frac{dx}{\sqrt{(a^4-x^4)}}\ \text{from}\ x=0\ \text{to}\ x=a,$$

$$(3) \quad \frac{\left(1 + \dfrac{dy^2}{dx^2}\right)^{3/2}}{\dfrac{d^2y}{dx^2}} = \frac{a^2}{2x},$$

$$(4) \quad \frac{dx}{\sqrt{(1-x^2)}} + \frac{dy}{\sqrt{(1-y^2)}} = 0,$$

$$(5) \quad xy - \frac{d^2z}{dx\, dy} = 0,$$

Gwatkin[17] followed with 'Integrate equations $\sqrt{(x)}\, dy = \sqrt{(y)}\, dx + \sqrt{(y)}\, dy$;

$$x\frac{dz}{dx} + y\frac{dz}{dy} = z\sqrt{x^2 + y^2},$$

but also included 'Find the fluxion of the arc whose tangent $= \sqrt{\dfrac{1-x}{1+x}}$,' and 'find the fluxion of the log of $\dfrac{x}{\sqrt{(1+x^2)}}$'. However, despite these references to fluxions, there is not a single dot to be found in any of the questions.

The fluxionists who had patiently taken such a battering from the younger men at the university now made their first and only reply in the controversy. A book was written at the University's expense by a retired member, Rev. D. M. Peacock, which strongly criticised the new methods and notation of analysis, the ideas of Lacroix on calculus, and the forceful way these had been imposed on the undergraduates in their examinations. The full title is *A Comparative View of The Principles of the Fluxional and Differential Calculus addressed to the University of Cambridge* and was published by the University Press in 1819. It is significant that the work did not appear until this date and shows how at a loss the reactionaries were to give any official and coherent reply to the work of the Analytical Society, with their publications of the *Memoirs* in 1813 and the translation of Lacroix in 1816, together with the teaching and examination work of George Peacock in 1817. The criticism of D. M. Peacock is powerfully expressed, however, and will be quoted at length as representing, in the absence of any other publication, the official answer of the university authorities to the members of the Analytical Society.

He commences:

> Though long retired from the University, the writer of the
> following pages has not been an unconcerned observer of its
> system of education; and it has been matter not less of surprise,
> than of regret to him, to see the success of the attempts that have
> been made of late years to introduce the Differential Calculus in
> the place of that of Fluxions. Had the Calculus been any new
> discovery, he could have ascribed its favourable reception to its
> novelty: but when he reflects that it is of the same date with the
> Newtonian system, that they have been rivals for a whole
> century, and that the Newtonian school has maintained its
> ground in this country through that long period, nay, that
> foreign analysts have found it necessary to approximate the
> principle of the Differential Calculus more and more nearly to
> that of Fluxions, he cannot but wonder at the eagerness shewn
> to explode a system so well sanctioned by experience; and still
> more at the success of these efforts.[18]

This is an extraordinary statement when the achievements
of Continental mathematicians such as Euler, Lagrange and
Laplace, using the differential calculus, are weighed against
the comparatively negligible discoveries of the British mathematicians in this century. D. M. Peacock then turns his attention
to the question of notation.

> As to the question, which system of notation should be adopted,
> the writer of these pages deems it a matter of very trifling
> importance, and he thinks that a most disproportionate degree
> of consequence has been assigned to it, and particularly by the
> translators of Lacroix's Elementary Treatise, who observe, that
> it constitutes 'the most important distinction' between the two
> methods. Were there no more important distinction between
> the two, he should hardly think the issue worth the struggle. He
> certainly thinks that the Fluxional notation ought to be
> retained, so far as it is found adequate, out of regard to the
> memory of the immortal Newton; but still he cannot but look
> upon it as a point of little consequence in itself, what symbols we
> use, so long as we agree in the ideas, that are to be annexed to
> them.[18]

This is the normal non-rational reason given for retaining
Newton's system, but further on the author offers a more critical
view.

> The Fluxional notation, he (the author) maintains, is in ordi-
> nary cases more compendious, inasmuch as $\dot{x}$ is plainly shorter

than dx. It has moreover the signal advantage of being specific, and thereby clearly distinguishing the fluxional part of a formula from the rest of it; whereas the symbol for a differential confounds itself with the coefficient, and is not otherwise distinguishable from it, than as the letter (d) is exclusively appropriated to denote it. But this is the least ambiguity attending the Differential notation; it affords no means of distinguishing the differentials of powers from the powers of differentials, but by the insertion of a dot, the nth power of dx being written dx^n, and the differential of x^n, d $\cdot$ x^n; and yet the latter distinction is obtained at the price of consistency, for according to the analogy of the latter notation, the differential of x ought to be written d $\cdot$ x and not dx; and were it so written, there would be no symbol left for the nth power of a differential; for the most obvious mode of expressing the latter and that, which the analogy of algebraic notation requires, is d$^n x$; but this is appropriated to express the differential of the nth order. In short, the differential scheme of notation is undoubtedly confused, and must necessarily be very perplexing to the student, as he has not only to learn a new set of associations, but to discard old ones, or more properly speaking, to learn to associate different ideas to the same symbols, according as he is reading a treatise on the Differential Calculus or one on Common Algebra; whereas the Fluxional notation is free from all ambiguity in the points just noticed, clearly distinguishing the fluxion of a power from the power of a fluxion, and the latter again from the higher orders of fluxions, and that agreeably to the common algebraic notation; and it may also be added, that it is not only fully adequate to all the ordinary forms and operations of the fluxional calculus, but by a little ingenuity might be easily extended, so as to express those which occur in the new methods of partial differences and variations.[19]

This point is well made, but the ambiguities he suggests can be overcome by a more developed form of the notation. His objections should be weighed against the sort of criticism of fluxionary notation made by Woodhouse in the *Principles of Analytical Calculation*, pp. xxvii–xxx.

It is curious that D. M. Peacock does not refer to this work by Woodhouse. He does mention Woodhouse twice, on pp. 80 and 83, in an approving way, but with reference to his treatise on *Isoperimetrical Problems*. It therefore seems extraordinary that D. M. Peacock should not refer to the *Principles of Analytical Calculation*, which is so relevant in that it takes a completely opposite view of the philosophy and notation of calculus and was written sixteen years earlier by a man he must have known and admired.

D. M. Peacock ends with an attack on those seeking to impose their views forcibly by way of the Cambridge examinations.

> The writer of these pages is as much disposed as any one to condemn a bigotted attachment to ancient systems, and that illiberal jealousy which would indiscriminately oppose the admission of all new matter that may be drawn from the works of a rival; though he still thinks, that a just distinction should be carefully made between what is really useful, and what is not; for it does not follow, because the French analysts have pushed the Differential Calculus farther than the Fluxional Calculus has been carried, that all the new matter they have added deserves to be adopted, or is fitted to be made matter of examination in the Senate-House. There is in truth no one point, on which the University should be more upon its guard, than against the introduction of mere algebraic or analytical speculations into its public examinations. This is continually objected to in the course of mathematical study pursued at Cambridge, and there is too much justice in the objection. Academical education should be strictly confined to subjects of real utility, and so far as the lucubrations of French analysts have no immediate bearing upon philosophy, they are as unfit subjects of academical examination, as the Aristotelian jargon of the old schools, nor is it right or reasonable, that young men should be obliged to read them in order to attain to the honours of the Senate-House. The different branches of Natural Philosophy, particularly Physical Astronomy, afford ample scope; and much of the time which young men are now obliged to bestow upon mere Algebra would be much more usefully devoted to these subjects. But whatever difference of opinion may prevail as to the propriety of limiting the discretion of the Moderators and Examiners as to the extent to which they shall introduce pure Algebra and Analytics into the examinations, it is to be presumed, that all will agree in this, that some control should be exercised as to the character of the books to be examined in, and that a Calculus founded on incontrovertible principles should not be exploded to make way for one radically defective.[20]

Thus D. M. Peacock condemns the new type of analysis introduced into the university examinations as being useless and unphilosophical. His castigation of this French analysis as having no immediate bearing on philosophy, coupled with his commendation that natural philosophy, especially physical astronomy, should be studied instead, indicates that he must have been unfamiliar with the work of Lagrange, Laplace and other Continental mathematicians who used the Leibnizian calculus so effectively in solving problems on astronomy and other branches of natural science.

George Peacock responded in a letter to Babbage, 23 November 1819:

> D. M. Peacock has written a furious pamphlet against the translators of Lacroix and particularly against me: though it shows some acuteness it is full of absurdities: as I am the principal object of the attack, I hope you will grant me the privilege of answering it. Tell Herschel of this, for he is attacked, though with much less asperity than myself.[21]

Unfortunately there is no record of George Peacock's actual reply. No further attacks on the method and notation of calculus were made. At this point in the history it seems that William Whewell played a major part in helping to establish the new symbolism in England. In an earlier letter of 1817 which I have quoted, he expressed his doubts about the success of G. Peacock's project, but in November, 1818, he was less cautious, for in a letter to Herschel, parts of which I have already quoted, he said, 'You see that I talk to you about these matters taking it for granted that you still retain some interest for your old plan of reforming the mathematics of the university. I have it now in my power to further this laudable object by the situation I have taken of assistant tutor (i.e. Mathematical Lecturer) here.'[22]

Whewell made two major contributions. In 1819 he published a book, *Elementary Treatise on Mechanics*, Cambridge, 1819, in which he consistently used the differential notation, the first English book on applied mathematics to do so, and then in 1820 was appointed moderator together with Henry Wilkinson. The 1820 examination was uncompromisingly in favour of the differential notation, and fluents, fluctions and dots were banished completely. George Peacock, as moderator again in 1821, continued in this vein, and from this time onwards the differential notation was used exclusively.

Meanwhile, what became of the Analytical Society? The Babbage correspondence indicates that a meeting was subsequently held at Thurlby Hall, Newark, the home of E. F. Bromhead:

At a Meeting of the Anal. Soc. held at Thurlby December 20, 1817.
Edward ffrench Bromhead Esq. in the Chair.
The Green M.S. being read it was unanimously resolved
1. That the Business of this Society shall for the future be transacted in London, in conformity with the Resolutions following, until further measures be taken for perfectly organising the Institution.
2. That Charles Babbage Esq., M.A., F.R.S., No. 5, Devonshire Street, Portland Place shall be Secretary.
3. That new Members shall be admitted upon the Recommendation of three Members, signified to the Secretary.
4. That Memoirs, Problems, Difficulties stated, and other Communications, shall be addressed to the Secretary, at whose house they may be seen, and by whom active steps will be taken for conveying information among the Members.
5. That the carriage of all Letters and Papers addressed to the Secretary shall be paid, and that no other expense be laid on the Members, until the Society shall be more extended and better established.
6. That the Secretary shall appoint a Day for an annual Meeting and Dinner of the Members, and shall appoint other meetings as occasions may require.
7. That Memoirs will be communicated *in the Name of the Society* to the Royal Society or other Literary Bodies, until it shall be deemed proper to publish the Next Number of the Society's own Transactions.
8. That these Resolutions be communicated to the present Members of the Society.[23]

The Society never did become more extended and better established, nor did it publish any further Transactions. The reason may be signified in a letter from George Peacock to Babbage two years later.

I was sure that you would rejoice at the establishment of a philosophical society at Cambridge. I send you a copy of the regulations, which have been agreed upon by the committee and which will be confirmed at the first meeting of the Society. I shall enter your name as a member.
It will supply that stimulus which Cambridge particularly wants and more than ever now, when the better system is almost universally adopted.[24]

It would appear that the Analytical Society had fulfilled its task, and the further needs of its members were met by the institution of the Cambridge Philosophical Society.

In conclusion it may be said that Woodhouse was certainly the first to introduce the Continental notation to England in his *Principles of Analytical Calculation*, but that his recommendations were rejected, or, what seems more likely, ignored. He was a great logician and had a forthright personality, but he made no apparent attempt to impose his views on the university or the country, as his younger contemporaries did a few years later. In fact his major work of 1803 appears to have been his only contribution to the debate, and all we know of its effect was that it introduced the Leibnizian notation to Babbage. The three great champions of the differential notation pursued their course independently and succeeded in airing their ideas by legitimate, but unorthodox, and highly effective means. In giving the credit to Peacock, Babbage and Herschel for their efforts which led quickly to a rebirth in British mathematics, freed from the shackles of traditionalism and misplaced patriotism, it is right to remember also the work of Woodhouse, who initiated the movement; of Ivory, who followed the continental methods independently; and of Whewell, who by his influence and publications finally established the change.

NOTES

1. C. Babbage, *Passages from the Life of a Philosopher*, London, 1864, p. 26.
2. *Ibid.*, p. 27.
3. *Ibid.*, p. 29.
4. *Memoirs of the Analytical Society*, Cambridge, 1813, p. i.
5. *Ibid.*, pp. ii–iii.
6. *Ibid.*, p. iv.
7. *Ibid.*, p. xvi.
8. Babbage, *op. cit.*, pp. 38–40.
9. W. W. R. Ball, *A History of the Study of Mathematics at Cambridge*, London, 1889, p. 193.
10. J. M. Stair Douglas, *The Life and Selections from the Correspondence of William Whewell D.D.*, London, 1881, p. 56.
11. *Cambridge Problems: being a collection of the printed questions proposed to the Candidates for the Degree of Batchelor of Arts at the General Examinations, from the year 1801 to the year 1820 inclusively*, London, 1836, pp. 93–9.
12. *Ibid.*, pp. 99–104.
13. I. Todhunter, *Life of Whewell*, London, 1876, vol. II, p. 16.
14. *Proceedings of the Royal Society*, 1859, **9**, 538.
15. Cambridge Problems, *op. cit.*, pp. 104–10.

16. Todhunter, *op. cit.*, p. 31.
17. Cambridge Problems, *op. cit.*, pp. 110–15.
18. Rev. D. M. Peacock, *A Comparative View of the Principles of the Fluxional and Differential Calculus addressed to the University of Cambridge*, Cambridge, 1819, pp. 1–3.
19. *Ibid.*, pp. 61–3.
20. *Ibid.*, pp. 84–6.
21. British Museum Additional Manuscripts 37182, No. 181.
22. Todhunter, *op. cit.*, pp. 29–31.
23. British Museum Additional Manuscripts 37182, No. 91.
24. *Ibid.*, No. 177.

4

The calculus of functions

The calculus of functions is undoubtedly Babbage's major mathematical invention. The subject, which we will see was taken up by Babbage almost from its origin and developed by a series of ingenious generalisations, has possibilities that have been little explored even in modern mathematics. The calculus of functions is developed on lines analogous to differential equations and difference equations, but is more general than either, and can be considered as including these areas as special cases.

The major ideas are presented in two papers published in the *Philosophical Transactions* of 1815 and 1816. There are references in other papers which are much shorter, but the two earliest ones are by far the most significant. Both are very long papers·and between them occupy 111 pages. Babbage did not develop the subject in his later years, but as he says of the calculus of functions in *Passages from the Life of a Philosopher*:

> This was my earliest step, and is still one to which I would willingly recur if other demands on my time permitted. Many years ago I recorded, in a small MS. volume, the facts, and also extracts of letters from Herschel, Bromhead and Maule, in which I believe I have done justice to my friends if not to myself. It is very remarkable that the Analytical Engine adapts itself with singular facility to the development and numerical working out of this vast department of analysis.[1]

The volume referred to may be found in the collection of Babbage relics contained in a cowhide trunk at the Museum of the History of Science, Oxford. The work consists of 289 closely written pages entitled 'The History of the Origin and Progress of the Calculus of Functions during the years 1809, 1810 . . . 1817'.

The manuscript has little mathematical content of any importance to add to Babbage's published work on the subject, but the personal details are interesting. He says that he began to think about the calculus of functions as early as the latter end of 1809. At this time he would have been only eighteen years old, and possibly seventeen if some biographers are right. This is remarkable in itself, as owing to sickness it appears he had no education at all before the age of ten, and apart from a few years' tuition in a school at Enfield, he had had very little formal mathematical teaching. His work on the calculus of functions must have begun during the period when he had left Enfield and was living at home in Totnes, with a classics tutor, before going to Cambridge in 1811.

It is also interesting to learn that he first came across a functional equation through attempting, like Descartes, to solve a geometrical problem of Pappus.

When he went to Cambridge, Babbage continued his work on the calculus of functions, and through the Analytical Society first met Herschel, Bromhead and Maule who also took a great interest in the subject. The four young mathematicians worked together during term time to solve some of the basic functional equations, and corresponded extensively during the vacations.

It is clear that Babbage did effective mathematical research not only in his undergraduate days, but even before going to Cambridge, always independent of any formal tuition or supervision. One cannot help wondering if a genius like his might be lost in a modern educational system.

The purpose of this chapter will be to consider: (*a*) the content of Babbage's work on the calculus of functions; (*b*) the extent to which we may claim originality for him in this field; and (*c*) the subsequent development and intrinsic value of this branch of mathematics.

The content is rather substantial but needs to be examined in some detail for a consideration of the variety of techniques used to solve problems, and the validity of the results.

The best place to start is Babbage's third published paper, 'An essay towards the calculus of functions, Part I' taken from *Philosophical Transactions*, 1815, **105**, 389–423. This is Babbage's first introduction of the subject and the first pages give the necessary notation and definitions.

He defines a function (*ibid.*, 389) as the result of every operation that can be performed on quantity. This is a wider definition than that given by many of his contemporaries who tried to restrict the idea of function to various combinations of prescribed algebraic manipulations. It is interesting that Babbage takes the function to be a general operation, and in the sense that the two terms are understood today it would probably be more accurate to describe his work as an 'operational' rather than a 'functional' calculus.

He observes on the same page that calculations with functions usually consist of two types, the direct and inverse ones, and that, of these two, as with the ordinary calculus, the inverse one is usually the more difficult and the more useful.

On p. 390 he describes the notation. The Greek letters α, β, γ, ... are taken to represent known functions and ψ, χ, ϕ, ... unknown ones. Throughout the work he uses ψx where we would write $\psi(x)$ and this takes some getting used to at first, but it should be realised that this is consistent with his notational innovation that when the same function operates twice successively on the variable x, the result is to be written $\psi^2 x$, and if n times, then as $\psi^n x$. If brackets were to be used for $\psi(x)$ then they should also be used for $\psi[\psi(x)]$, and the result would be a clutter of bracket symbols when higher orders are considered. However, the meaning of $\psi^n x$ is quite unambiguous and all the brackets can be eliminated.

There is a theory of mathematical invention that a new subject begins when a satisfactory notation is first given. In calculus many problems connected with finding tangents and maxima and minima were solved by various apparently independent methods, but when Newton provided the 'dot' and Leibniz the 'd' notation, these methods were seen as part of one general subject which could now develop much more rapidly. In the same way we will see how dependent the development of the calculus of functions is on this new functional notation.

A more difficult notation, introduced on p. 391, reads $\psi^{2,1}(x, y)$ for $\psi[\psi(x, y), y]$ and $\psi^{1,2}(x, y)$ for $\psi[x, \psi(x, y)]$. Equations involving functions of this type are called functional equations of two variables. Babbage claims, on the same page, that solutions to first order equations have been known, but that his more general approach is original, and that previously second and higher order equations had not even been mentioned. The first order functional equations, solved by others, he considers (*ibid.*, 393) most probably arose from the integration of partial differential equations, and he names D'Alembert, Euler, Lagrange, Monge, Laplace, Arbogast and Herschel. I will consider their work in this field later.

Before attempting any problems he distinguishes between particular and general solutions. One of the difficulties of this work is to know how general a general solution is, but he takes it for the moment to be a solution which contains one or more arbitrary functions. We find in nearly all the problems that the usual technique is to take a particular solution of a problem for granted and to generalise from there. A particular solution can often be found easily enough by guesswork, but the problem does arise as to whether the general solution obtained in this way is general enough.

Before looking at any of the problems, it will be useful to write down a list of those tackled in this paper, in order to see the development from the simple to the more complex. The functional equations solved, sometimes by alternative methods, are in order.

1. $\psi x = \psi \alpha x.$
2. $\psi x = Ax \cdot \psi \alpha x,$
3. $\psi x = Ax \cdot \psi \alpha x + Bx.$
4. $\psi x + Ax \cdot \psi \alpha x + \cdots + Nx \cdot \psi \nu x + X = 0$
5. $\phi x = \phi \alpha x = \phi \beta x = \cdots = \phi \nu x.$
6. $\psi \alpha x = \psi \alpha_1 x; \ \psi \beta x = \psi \beta_1 x' \ldots \psi \nu x = \psi \nu_1 x.$
7. $F(x, \psi x, \psi \alpha x, \ldots \psi \nu x) = 0.$

$$\left\{ \begin{array}{l} \alpha, \beta, \ldots \nu, \\ \alpha_1, \beta_1, \ldots \nu_1, \\ A, B, \ldots N, X \text{ are} \\ \text{known functions.} \end{array} \right.$$

These are followed by some second and higher order problems:

8. $\psi^2 x = x.$
9. $\psi^n x = x.$
10. $\psi^2 x = \alpha x.$
11. $\psi^n x = \alpha x.$
12. $\psi \alpha \psi \beta x = \gamma x.$
13. $\psi \alpha (y, \psi y) = y.$
14. $F(x, \psi x, \psi^2 x, \ldots \psi^n x) = 0.$
15. $F(x, \psi x, \psi^2 \alpha x, \ldots \psi^n \nu x) = 0.$
16. $F(x, \psi x, \psi \alpha (x, \psi \beta x)) = 0.$

We shall see that the aim is to solve functional equations of a very general nature. The methods which are used are almost as interesting as the solutions themselves and often show analogies with devices used to resolve differential and difference equations.

Babbage develops this theme of invention by analogy in his unpublished work 'The Philosophy of Analysis' which will be examined in chapter 5.

Problem I is on p. 395. It is to find the general solution of the equation $\psi x = \psi \alpha x$, given a particular solution. The solution is very simple. If a function which satisfies the equation is f, then $fx = f\alpha x$. If ϕ is any function, then $\phi fx = \phi f\alpha x$. This means that $\psi = \phi f$ is a general solution, containing the arbitrary function ϕ.

How general is it? This question is well illustrated by the example which follows.

Take $\alpha x = -x$, so that the equation becomes $\psi(x) = \psi(-x)$. A solution of this is $f(x) = x^2$. Then, the general solution is $\phi(x^2)$ where ϕ is any function. It will certainly be true that any rational function of the form $\phi(x^2)$ satisfies the equation $\psi(x) = \psi(-x)$, but one is left wondering if there are not other solutions of a different form. Any even function, by definition, satisfies $\psi(x) = \psi(-x)$ but it is not obvious that all of these are of the form $\phi(x^2)$. There are for example the functions $\cos x$ and $\cosh x$ which can be considered as power series in x^2, and if $\phi(x^2)$ can include infinite series there is no difficulty. However, we cannot allow functions like $\phi(x) = x^{3/2}$, for then $\phi(x^2) = (x^2)^{3/2} = x^3$ which most certainly does not satisfy $\psi(x) = \psi(-x)$ since $\psi(x) = x^3$ and $\psi(-x) = -x^3$. We could argue somewhat artificially, that $(x^2)^{3/2}$ can be considered an even function by inserting that x be replaced by $-x$ before the exponentiation is performed. It would seem better to restrict ϕ to the class of rational functions where no problem arises. It may be that this was an unfortunate special case to take, and if we consider instead the equation $\psi(x) = \psi(1-x)$, then a solution is given by $f(x) = x(1-x)$, and the general solution $\phi[x(1-x)]$ allows ϕ to be any type of function.

On p. 396, a new notation is introduced. The symbol $\bar{\phi}(x, y)$ is taken to mean a symmetrical function in x and y so that $\bar{\phi}(x, y) = \bar{\phi}(y, x)$. This is used to tackle the same problem, $\psi x = \psi \alpha x$ by an alternative method. Suppose we have instead of a particular solution for this equation, the solution of $fx = f\alpha^2 x$.

Assume now (*ibid.*, 397) that $\psi x = \bar{\phi}(fx, f_1 x)$ where $f_1 x$ is another function, to be determined.

Then since $\psi x = \psi \alpha x$, $\bar{\phi}(fx, f_1 x) = \bar{\phi}(f\alpha x, f_1 \alpha x)$. As the functions are both symmetrical, f_1 can be found by solving $fx = f_1 \alpha x$ and $f_1 x = f\alpha x$. But since $f_1 x = f\alpha x$, $f_1 \alpha x = f\alpha^2 x$, replacing x by αx, and $fx = f\alpha^2 x$. But fx was defined to satisfy this equation, and by taking $fx = f\alpha x$, a solution of the original equation is obtained. The general solution is, therefore, $\psi x = \bar{\phi}(fx, f\alpha x)$.

Now if α has the property that $\alpha^2 x = x$, the equation $fx = f\alpha^2 x$ will be satisfied by any function f, and so in particular by $fx = x$. This gives an immediate solution to a problem when α has this nature. If, in particular,

$$\alpha x = \frac{x}{ax - 1},$$

then

$$\alpha^2 x = \frac{\dfrac{x}{ax - 1}}{a\left(\dfrac{x}{ax - 1}\right) - 1} = x,$$

and the equation $\psi x = \psi[x/(ax - 1)]$ has the general solution

$$\psi x = \bar{\phi}\left(x, \frac{x}{ax - 1}\right).$$

When $a = 0$, this gives equation $\psi x = \psi(-x)$ with solution $\psi x = \bar{\phi}(x, -x) = \phi(x^2)$ as before.

Problem III (*ibid.*, 398) is to solve $\psi x = Ax \cdot \psi\alpha x$. Two solutions are given, the first by a method similar to the previous one, and the second by a new one which will therefore be the more interesting case. To solve $\psi x = Ax \cdot \psi\alpha x$, suppose as usual there is a particular solution fx.

Put $\psi x = fx \cdot \phi x$. Then

$$fx\phi x = Ax \cdot f\alpha x \cdot \phi\alpha x. \tag{1}$$

But since fx satisfies the original equation, $fx = Ax \cdot f\alpha x$, and (1) becomes $\phi x = \phi\alpha x$, which is a type already solved. If $f_1 x$ is a particular solution of this, $\phi f_1 x$ is the general solution and so the solution to the general equation is $\psi x = fx \cdot \phi f_1 x$.

As an example, Babbage considers the equation $\psi x = [1 - x^2/(1 - 2x)]\psi[x/(x - 1)]$ and obtains the general solution $\psi x = (1 + x)\phi[x^2/(x - 1)]$. From this result he adds (*ibid.*, 398) that another solution is $\psi x = (1 + x)\phi[\cos \log (x - 1)]$, indicating that this general function ϕ could range quite freely through operations within infinite series.

Problem IV (*ibid.*, 400) considers $\psi x = Ax \cdot \psi \alpha x + Bx$. The technique here is to take a particular solution fx, let $\psi x = fx + \phi x$, and then the problem reduces to the previous type. This method and the one used in Problem III have their obvious counterparts in differential equation theory.

The same technique of adding an arbitrary function to a particular solution is used again in Problem V (p. 402) to reduce the equation $\psi x + Ax \cdot \psi \alpha x + \cdots + Nx \cdot \psi \nu x + X = 0$ to one in which the last term X is eliminated.

Problem VI (p. 403) is a different type. It is to find a function of x such that if $\alpha x, \beta x, \gamma x, \ldots$ are substituted in turn for x, the function remains unchanged. This means that ϕx is to be found satisfying the equations $\phi x = \phi \alpha x = \phi \beta x = \cdots = \phi \nu x$. ϕ is found by a constructive method. First find any f such that f$x =$ fαx. Then with this particular f substituted in, find any f_1 which satisfies $f_1 fx = f_1 f\alpha \beta x = f_1 f\beta x$. Then find an f_2 such that $f_2 f_1 fx = f_2 f_1 f\beta \gamma x = f_2 f_1 f\gamma x$, and continuing the process, finally,

$$f_n f_{n-1} \ldots fx = f_n f_{n-1} \ldots f\nu x.$$

Then if ϕ is arbitrary, $\phi\{f_n f_{n-1} \cdots f_1 fx\}$ satisfies the conditions.

This method is used in an example to find a function which remains the same when x, $-x$ and $x/\sqrt{(x^2 - 1)}$ are substituted for x. The result found is $\phi[x^4/(x^2 - 1)]$.

A similar construction is used in Problem VII (*ibid.*, p. 405) to find a function ψ satisfying the simultaneous functional equations $\psi \alpha x = \psi \alpha_1 x$, $\psi \beta x = \psi \beta_1 x, \ldots \psi \nu x = \psi \nu_1 x$, with solution $\psi x = \phi f_n f_{n-1} \ldots f\alpha \beta \gamma \ldots \nu x$, where f, $f_1, \ldots f_n$ are successive particular solutions. At this stage Babbage asserts, 'We now possess the means of solving a much more general problem than we have yet attempted, it is the general solution of any functional equation of the first order' (*ibid.*, p. 406).

The general functional equation of the first order is $F(x, \psi x, \psi \alpha x, \ldots \psi \nu x) = 0$ (Problem VIII, *ibid.*, p. 406), and Babbage shows that this can be solved, given a particular solution which contains one or more arbitrary constants. The more such constants that can be found, the more general will be the solution.

Suppose there is an arbitrary function ϕx which remains unchanged when $\alpha x, \beta x, \ldots \nu x$ are substituted for x in turn. By Problem VI, such a function can be constructed. Let the particular solution be f$(x, a, b, c, \ldots)$ where $a, b, c, \ldots$ are the arbitrary constants.

Then on substitution, the original equation becomes

$$F[x, f(x, a, b, \ldots), f(\alpha x, a, b, \ldots), \ldots f(\nu x, a, b, \ldots)] = 0$$

and the $a, b, c, \ldots$ must all disappear identically.

Now consider the expression $f(x, \phi x, \phi_1 x, \phi_2 x, \ldots)$ where ϕx, $\phi_1 x, \phi_2 x, \ldots$ are all functions remaining unchanged when x is replaced in turn by $\alpha x, \beta x, \ldots \nu x$.

Then if

$$\psi x = f(x, \phi x, \phi_1 x, \ldots)$$

we have

$$\psi \alpha x = f(\alpha x, \phi x, \phi_1 x, \ldots)$$
$$\psi \beta x = f(\beta x, \phi x, \phi_1 x, \ldots)$$
$$\vdots \qquad\qquad \vdots$$
$$\psi \nu x = f(\nu x, \phi x, \phi_1 x, \ldots)$$

and on substitution in our original equation,

$$F(x, \psi x, \psi \alpha x, \ldots \psi \nu x)$$
$$= F[x, f(x, \phi x, \ldots), f(\alpha x, \phi x, \ldots), \ldots f(\nu x, \phi x, \ldots)] = 0,$$

so that the $\phi x, \phi_1 x, \ldots$ all disappear in the same way that $a, b, c, \ldots$ did. Thus $\psi x = f(x, \phi x, \phi_1 x, \ldots)$ is a general solution with as many arbitrary functions as the particular solution had arbitrary constants.

It is necessary to pause in admiration of a method so simple and elegant for solving such a general problem. Providing a particular solution is known with at least one arbitrary constant, a general solution may be found to any functional equation of the first order.

Babbage's illustrative example (*ibid.*, p. 407) is $x\psi x = \psi(1/x)$. A particular solution is $\psi x = a[(x+1)/x]$ where a is an arbitrary constant.

$\bar{\phi}[x, (1/x)]$ is a function which remains unchanged when x is replaced by $1/x$, and this completes the general solution $\psi x = [(x+1)/x]\bar{\phi}[x, (1/x)]$. This general solution with its apparently 'arbitrary' number of arbitrary functions leads immediately to the question of how many there ought to be. Babbage now considers this difficult problem, as he puts it, 'On the number of arbitrary functions introduced into the complete solution of a functional equation' (*ibid.*, p. 408).

He continues on the same page:

> The train of reasoning usually made use of to prove, that a differential equation of the *n*th order, requires in its complete integral *n* arbitrary constants may be pursued on the present occasion, though from several reasons, it would perhaps be desirable to have a proof resting on a different principle; as I have not been successful in discovering any other. I shall give the only one I am at present possessed of.

The proof he gives is a brave try but seems to work at the problem from the wrong end and clearly does not overcome his own doubts.

We begin with a solution containing n arbitrary constants and try to find what sort of functional equation could give rise to one of this nature. Taking the solution as $\psi x = F(x, a_1, a_2, \ldots a_n)$, replace x in turn by any number of known functions αx, $\beta x, \ldots \nu x$.

The results will be

$$\psi x = F(x, a_1, \ldots a_n) \tag{0}$$

$$\psi \alpha x = F(\alpha x, a_1, \ldots a_n), \tag{1}$$

$$\vdots \qquad \vdots$$

$$\psi \nu x = F(\nu x, a_1, \ldots a_n). \tag{n}$$

From these $n + 1$ equations, eliminate the n arbitrary constants and the resulting equation will be of the form

$$0 = F_1(x, \psi x, \psi \alpha x, \ldots \psi \nu x), \tag{A}$$

where $\alpha, \beta, \ldots \nu$ are exactly n known functions.

If the converse of this result could be proved, the required result would be established. The argument itself is unfortunately irreversible, and to assume that a general equation like (A) had a general solution containing exactly n arbitrary constants (and, therefore, by the previous proof, the same number of arbitrary functions) is, according to Babbage, 'too hasty a conclusion'. He adds, on p. 409, 'The reasoning is certainly plausible, and such a solution is undoubtedly a very general one; still, however, there are reasons which incline me to believe, that other solutions exist of a yet more general nature.'

We can agree with him about this last statement. He does not indicate what the reasons were which led him to the idea that more general solutions existed, but we can strongly suspect that they do exist, without being able to supply a proof.

On p. 409, Babbage begins the section: 'On functional equations of the second and higher orders'. To solve such problems, new techniques are required, for 'The artifices which were used with success in the preceding part of this paper, are no longer of any avail.'

The first problem attempted is to find a function ψ such that $\psi^2 x = x$. Various methods are given, but the technique most useful to this and the subsequent problems is first given in Problem X, solved on p. 411.

The solution is obtained in terms of a particular solution by assuming that $\psi x = \phi^{-1} f \phi x$.
Then

$$\psi^2 x = (\phi^{-1} f \phi)(\phi^{-1} f \phi x),$$

$$= \phi^{-1} f \phi \phi^{-1} f \phi x,$$

$$= \phi^{-1} f f \phi x, \text{ since } \phi \phi^{-1} \text{ is the identical operation,}$$

$$= \phi^{-1} f^2 \phi x = x.$$

If now the function $f x$ is a particular solution of $\psi^2 x = x$, $f^2 x = x$ and $f^2 \phi x = \phi x$ so that the equation becomes $\phi^{-1} \phi x = x$, which is true since ϕ^{-1} is the inverse function of ϕ. This means that if f is a particular solution of $\psi^2 x = x$, then $\psi = \phi^{-1} f \phi$ is a general solution, ϕ being as usual an arbitrary function.

Exactly the same artifice is next used to solve the more general $\psi^n x = x$. It can easily be seen that if f is a particular solution, $\psi = \phi^{-1} f \phi$ will be the general one.

This is used again in the next two problems, the second (Problem XIII on pp. 413–14) being the more general. It is $\psi^n x = \alpha x$. Put $\psi x = \phi^{-1} f \phi x$, and, on substitution, $\phi^{-1} f^n \phi x = \alpha x$; therefore $f^n \phi x = \phi \alpha x$. But since f is a known function, f^n can be determined, and the problem reduces to a first order one.

A different method is used in Problem XV (written as XIV by a misprint) on p. 414. This is another attempt to solve $\psi^2 x = x$, and begins in the usual way with $\psi = \phi^{-1} f \phi$, leading to $\phi^{-1} f^2 \phi x = x$. Now let $y = \phi x$ so that $x = \phi^{-1} y$. Substituting for x this becomes $\phi^{-1} f^2 y = \phi^{-1} y$. But since f^2 is known, this is a first order equation in ϕ^{-1}, with solution $\phi^{-1}y = \bar{\chi}(y, f^2 y)$ from Problem II. According to Babbage, 'this method is much more extensive in its application than any of those before it' (*ibid.*, p. 415).

The method is used on pp. 416–20 to solve the general functional equation $F(x, \psi x, \psi^2 \alpha x, \ldots \psi^n \gamma x) = 0$.

This is first reduced to the form $F(x, \psi x, \psi^2 x, \ldots \psi^n x) = 0$, by taking $\psi x = A^{-1} \phi A x$.

Then

$$F(x, \psi x, \psi^2 \alpha x, \ldots \psi^n \gamma x)$$
$$\rightarrow F(x, A^{-1} \phi A x, A^{-1} \phi^2 A \alpha x, \ldots A^{-1} \phi^n A \gamma x) = 0.$$

A function Ax can be chosen which remains unchanged by the successive substitution $\alpha x, \beta x, \ldots \gamma x$ (Problem VI) and the equation reduces to $F(x, A^{-1} \phi A x, A^{-1} \phi^2 A x, \ldots A^{-1} \phi^n A x) = 0$. Babbage could have said at this stage that since $\psi^n = A^{-1} \phi^n A$, this equation has the required form, but curiously he chose to put y for Ax, obtaining $F(A^{-1}y, A^{-1} \phi y, \ldots A^{-1} \phi^n y) = 0$ which he claims erroneously is of the right form.

To solve $F(x, \psi x, \psi^2 x, \ldots \psi^n x) = 0$, take $\psi = \phi^{-1} f \phi$ so that the equation becomes $F(x, \phi^{-1} f \phi x, \ldots \phi^{-1} f^n \phi x) = 0$. Using the new technique, replace x by $\phi^{-1}x$. Then $F(\phi^{-1}x, \phi^{-1}fx, \phi^{-1}f^2x, \ldots \phi^{-1}f^n x) = 0$ which is an equation of the first order relative to ϕ^{-1} and can be solved by Problem VIII.

In problems of functional equations of second and higher orders, the question of the number of arbitrary functions in the general solution becomes very complex. Babbage admits:

> With respect to the number of arbitrary functions which enter into the complete solution of functional equations of higher orders than the first, I have little at present to offer; the difficulty of the subject, and the wide extent of the enquiries to which it would lead, induce me to postpone it until I have more time for the consideration. (*Ibid.*, pp. 420–1.)

The postponement proved to be indefinite. We can sympathise, as this is clearly a very difficult problem. As a postscript to the paper he adds:

> Since the above was written, I have bestowed some attention on functional equations involving two or more variables, and I have met with considerable success: I am in possession of methods which give the general solution of equations of all orders, and even of those which contain symmetrical functions.

> I have also discovered a new and direct method of treating
> functional equations of the first order, and of any number of
> variables, and this new method I have applied to the solution of
> differential and even of partial differential functional equa-
> tions. (*Ibid.*, p. 423.)

These more complicated types of problem are tackled in the
second part of 'An essay towards the calculus of functions'.

Before considering this, there are two interesting letters to
Babbage about the first paper. One is from E. F. Bromhead,
dated 9 February 1816, and indicates the reaction to such a paper
in those times:

> Very many thanks for your kind letter and invaluable Paper,
> which will bear a hundred readings. I had a long talk about it
> with Woodhouse who had not seen, but instantly ordered it on
> my very strong recommendation. Others spoke about it, but I
> did not find a single Soul, even among Senior Wranglers,
> Herschel excepted, who understood a word about it. One very
> high man a fellow, observed that $\phi\alpha(x)$ could not $= \phi(x)$, as then
> $\alpha(x) = x$. Q.E.D.[2]

It is amusing to see again the contrast between the Analytical
Society plus Woodhouse and the other mathematicians at Cambridge. One can have a certain amount of sympathy with 'the very
high man', in that the rather loose and undefined way that the
inverse operator ϕ^{-1} was used in the paper might well tempt one
to apply it with such drastic results to the equation $\phi\alpha(x) = \phi(x)$.

The other letter contains a mathematical suggestion and criticism from James Ivory, dated 21 February 1816.[3]

> Your equation $\psi^2 x = x$, may thus be solved.
>
> Assume $\psi x = x\phi x$; then $\psi^2 x = x\phi x \cdot \phi(x\phi x)$; therefore
> $\phi x \cdot \phi(x\phi x) = 1$; and, by putting $\phi^{-1}x$ for x, we get
> $x \cdot \phi(x\phi^{-1}x) = 1$: therefore
>
> $$\phi(x\phi^{-1}x) = \frac{1}{x};$$
>
> and by taking ϕ^{-1} on both sides,
>
> $$x\phi^{-1}x = \phi^{-1}\frac{1}{x}:$$
>
> therefore
>
> $$x^{1/2}\phi^{-1}x = \left(\frac{1}{x}\right)^{1/2}\phi^{-1}\left(\frac{1}{x}\right)$$

the equation is now under the form

$$fx = f\left(\frac{1}{x}\right),$$

and therefore

$$x^{1/2}\phi^{-1}x = f\left\{\bar{x}, \frac{\bar{1}}{x}\right\} \quad \text{and} \quad \phi^{-1}x = \frac{1}{\sqrt{x}} \times f\left\{\bar{x}, \frac{\bar{1}}{x}\right\}.$$

This analysis is quite general and it does not require a case.

The argument seems a fair one, and a skilful manipulation of this calculus. Ivory continues:[3]

> As much depends on the solution of the equation $\psi x = \psi\alpha x$, it ought to be clearly made out. You demand a law, which is not altogether satisfactory. In the given equation put αx for x, then $\psi\alpha x = \psi\alpha^2 x = \psi x$; and again putting αx for x, in $\psi x = \psi\alpha^2 x$, we get $\psi x = \psi\alpha^3 x$; therefore if $\psi x = \psi\alpha x$, we must have $\psi x = \psi\alpha x = \psi\alpha^2 x = \psi\alpha^3 x$ &c. I conclude therefore that the equation $\psi x = \psi\alpha x$ cannot be solved unless $\psi x = \phi\{\bar{x}, \overline{\alpha x}, \overline{\alpha^2 x}, \ldots \overline{\alpha^{n-1}x}\}$ and $\alpha^n x = x$.

The reasoning again seems perfectly valid. Babbage had found that it is certainly much easier to obtain a solution when there is an n such that $\alpha^n x = x$, and the examples he gives are of this type exclusively. However, he never expressed this as a general condition, and Ivory is right to say that this must hold before any solution is possible.

We now turn to the second essay on this work, 'An essay towards the calculus of functions, Part II', *Philosophical Transactions*, 1816, **106**, 179–256. The paper was communicated by W. H. Wollaston and read on 14 March 1816. This is a long, complicated and ambitious paper. Babbage asserts at the beginning that his discoveries here will not only be of benefit to pure mathematics, but are highly useful to many branches of natural science.

> Nor is it only in the recesses of this abstract science, that its advantages will be felt: it is peculiarly adapted to the discovery of these laws of action by which one particle of matter attracts or repels another of the same or of a different species; consequently, it may be applied to every branch of natural philosophy, where the object is to discover by calculation from the results of experiment, the laws which regulate the action of ultimate particles of bodies...

But should the labours of future enquirers give to it that perfection, which other methods of investigation have attained, it is not too much to hope, that its maturer age shall unveil the hidden laws which govern the phenomena of magnetic, electric, or even of chemical action. (*Ibid.*, pp. 197–80.)

These prophetic statements have not been fulfilled but it may be that there are some results, especially towards the end of this paper, which have been overlooked by subsequent mathematicians.

It will be useful again to look first at the type of problem considered in the paper. There seem to be three major sections. The first contains problems in two variables, analogous to those studied in Part I. They are

$$\psi(x, y) = \psi(\alpha x, \beta y),$$

$$\psi(x, y) = \psi[\alpha(x, y), \beta(x, y)],$$

$$\psi(x, y) = A(x, y)\psi[\alpha(x, y), \beta(x, y)],$$

$$\psi(x, y) = A(x, y)\psi[\alpha(x, y), \beta(x, y)] + B(x, y),$$

$$\psi(x, y) = \psi[\alpha(x, y), \beta(x, y)] = \psi[\alpha_1(x, y), \beta_1(x, y)] = \cdots$$

The second group contains functional equations of the second and higher orders in two variables. These include

$$\psi^{2,1}(x, y) = x,$$

$$\psi^{\overline{2,2}}(x, y) = 0,$$

$$\psi^{\overline{2,2}}(x, y) = a,$$

$$\psi^{\overline{n,n}}(x, y) = ax + by,$$

$$\psi^{\overline{n,n}}(x, y) = a[\psi(x, y)]^b,$$

$$\psi^{\overline{\overline{2,2}}}(x, y) = F\psi(x, y),$$

$$\psi^{2,1}(x, y) = \psi^{1,2}(x, y),$$

$$x\psi^{1,2}(x, y) = y\psi^{2,1}(x, y).$$

The third group is the most interesting, for this contains examples of differential and integral functional equations of the first and second order in one and two variables. These are

$$\psi \alpha x = \frac{d\psi x}{dx},$$

$$\psi \alpha x = \frac{d^n \psi x}{dx^n},$$

$$F\left(x, \psi x, \psi \alpha x, \frac{d\psi x}{dx}\right) = 0,$$

$$\psi^2 x = \frac{d\psi x}{dx},$$

$$\int dx \psi^2 x = \psi x,$$

$$\psi(x, y) = \frac{d\psi(x, \alpha y)}{dx}.$$

We need not linger over the first section. The results are not very interesting, are not used to solve higher problems and the methods are almost identical to those given in the analogous cases in the first part.

$\psi(x, y) = \psi(\alpha x, \beta x)$ is solved by the simple stratagem of putting $\psi(x, y) = \phi(fx, f_1 y)$ (*ibid.*, p. 184), leading to a determination of f and f_1 from the equations $fx = f\alpha x$, $f_1 y = f_1 \beta y$. If f and f_1 are any particular solutions of these, then the general solution is given as $\psi(x, y) = \phi(fx, f_1 y)$ where ϕ is an arbitrary function.

To solve $\psi(x, y) = \psi[\alpha(x, y), \beta(x, y)]$ (*ibid.*, p. 187), it is assumed that $\psi(x, y) = \phi[f(x, y), f_1(x, y)]$. Substitution leads to

$$f(x, y) = f[\alpha(x, y), \beta(x, y)]$$

and

$$f_1(x, y) = f_1[\alpha(x, y), \beta(x, y)],$$

so that f and f_1 are particular solutions of the original equation. Thus if f and f_1 are any two particular solutions of $\psi(x, y) = \psi[\alpha(x, y), \beta(x, y)]$ the general solution is given by $\psi(x, y) = \phi[f(x, y), f_1(x, y)]$ where ϕ is an arbitrary function. The question of generality again arises, and Babbage says, 'Though these solutions may with propriety be termed general because they contain an arbitrary function, yet I am by no means inclined to think them the most general of which the questions admit.' (*Ibid.*, p. 189.)

The problem $\psi(x, y) = A(x, y)\psi[\alpha(x, y), \beta(x, y)] + B(x, y)$ (*ibid.*, p. 194) is reduced by the substitution $\psi(x, y) = f(x, y) + \phi(x, y)$ where $f(x, y)$ is a particular solution to a type of the form $\psi(x, y) = A(x, y)\psi[\alpha(x, y), \beta(x, y)]$. This is solved by a method analogous to the corresponding problem in the first paper, putting $\psi(x, y) = f(x, y)\phi[f_1(x, y), f_2(x, y)]$.

The most significant part of this section occurs on pp. 190–1, where Babbage gives his first general discussion on the nature of inverse operations. He takes the direct operation as being always single-valued and points out that the inverse function is then usually many-valued. The problem arises as to which of these several types of function to take so that the inverse operator becomes the unique one. The answer is that as for a given ϕ, ϕ^{-1} is defined by $\phi^{-1}\phi x = x$, then we must take the particular function ϕ^{-1} as the one which satisfies also $\phi\phi^{-1}x = x$. An example, given by Babbage, makes this quite clear. Suppose $z = \phi x = a - x^2$, which is single-valued then $x = \phi^{-1}z = \pm\sqrt{(a - z)}$, which is double-valued. Of these two possibilities, if we take $\phi^{-1}z = \sqrt{(a - z)}$, then $\phi\phi^{-1}x = \phi[\sqrt{(a - x)}] = a - (a - x) = x$. If on the other hand, $\phi^{-1}z = -\sqrt{(a - z)}$, then $\phi\phi^{-1}x = -x$. This is surely a misprint, for then $\phi\phi^{-1}x = \phi[-\sqrt{(a - x)}] = a - (a - x) = x$ as before. What must be meant is that $\phi^{-1}\phi x = \phi^{-1}(a - x^2) = -x$, and in the other case, taking $\phi^{-1}z = \sqrt{(a - z)}$, then $\phi^{-1}\phi x = x$. The question of inverse operators becomes more acute when functions of a more complicated nature are considered, but throughout this work on the calculus of functions, their existence is always assumed. One can make the point here that questions of continuity and differentiability should be considered in this context, but the general trend in Babbage's arguments can be followed without, at this stage, a too detailed discussion of existence theorems. At any rate his remarks on inverse functions were somewhat overdue in view of the use he had already made of them in both papers so far: 'This remark, which is of some importance, extends to the conclusions in my former Paper and to the whole of the subsequent enquiries' (*ibid.*, p. 191).

The next part of the paper is to solve problems: 'Of functional equations of the second and higher orders involving two or more variables' (*ibid.*, p. 197). These are very curious equations which quite probably have a value in mathematics where functions of several variables are dealt with in an iterative manner. The first problem of this type (Problem VIII, *ibid.*, p. 197) is $\psi^{2,1}(x, y) = x$. In view of the very compact notation used, this means $\psi[\psi(x, y), y] = x$.

The solution is again ingenious. Treat y as a constant, so that the equation becomes effectively $\phi^2 x = x$, find a particular solution, and replace any arbitrary constant in this solution by an arbitrary function of y.

For example, $f(x) = (b - x)/(1 - cx)$ satisfies $\phi^2 x = x$. Babbage then replaces the c by $\chi(y)$, an arbitrary function of y, obtaining $(b - x)/[1 - x\chi(y)]$ which can be seen to be a 'general' solution of $\psi^{2,1}(x, y) - x$. It could be added that the solution could be made more general by replacing the b by another arbitrary function $\phi(y)$ and then $[\phi(y) - x]/[1 - x\chi(y)]$ also satisfies the equation.

The next problems involve symmetrical substitutions $\psi^{\overline{n,n}}(x, y)$. The first of these is to solve $\psi^{\overline{2,2}}(x, y) = 0$, meaning $\psi[\psi(x, y), \psi(x, y)] = 0$. The function $x - y$ is obviously a solution, and so also is $(x - y)\phi(x, y)$, where ϕ is an arbitrary function. The same method is used to tackle $\psi^{\overline{2,2}}(x, y) = a$, where $\psi(x, y) = x - y + a$ is found as a particular solution and $(x - y)\phi(x, y) + a$ as a more general one. Taking the more general $\psi^{\overline{n,n}}(x, y)$, particular solutions are now sought, involving homogeneous functions.

The first of these Problem XI (*ibid.*, p. 201), the equation $\psi^{\overline{n,n}}(x, y) = ax + by$.

To find a particular solution, let $\psi(x, y) = px + qy$. Then

$$\psi^{\overline{2,2}}(x, y) = (p + q)(px + qy)$$

and

$$\psi^{\overline{n,n}}(x, y) = (p + q)^{n-1}(px + qy).$$

A solution can therefore be found by putting $a = p(p + q)^{n-1}$ and $b = q(p + q)^{n-1}$. Solving these equations for p and q,

$$p = \frac{a}{(a + b)^{\frac{n-1}{n}}} \quad \text{and} \quad q = \frac{b}{(a + b)^{\frac{n-1}{n}}}.$$

From this particular solution, a general solution $(px + qy)\phi(x, y)$ may be found.

On p. 202, Babbage makes the general assertion that if $\psi(x, y)$ is homogeneous in x and y (presumably of degree n), then

$$\psi^{\overline{k,k}}(x, y) = [\psi(x, y)]^{n^{k-1}}[\psi(1, 1)]^{\frac{1-n^{k-1}}{1-n}}.$$

Assuming that $\psi(x, y)$ is a polynomial, this result can be proved quite easily by induction.

This general result on homogeneous functions can now be used to solve a variety of problems, as for example, $\psi^{\overline{n,n}}(x, y) = a[\psi(x, y)]^b$ where appropriate values for a and b may be found.

After a series of problems arising from the applications of these theorems, the next type considered is typified by Problem XX (*ibid.*, p. 216) which is to solve the equation $\psi^{2,1}(x, y) = \psi^{1,2}(x, y)$ or $\psi[\psi(x, y), y] = \psi[x, \psi(x, y)]$. To solve this it is supposed that $\psi(x, y) = \phi^{-1}f(\phi x, \phi y)$. Then

$$
\begin{aligned}
\psi^{2,1}(x, y) &= \psi[\phi^{-1}f(\phi x, \phi y), y], \\
&= \phi^{-1}f(\phi\phi^{-1}f(\phi x, \phi y), \phi y), \\
&= \phi^{-1}f[f(\phi x, \phi y), \phi y], \\
&= \phi^{-1}f^{2,1}(\phi x, \phi y).
\end{aligned}
$$

Similarly, $\psi^{1,2}(x, y) = \phi^{-1}f^{1,2}(\phi x, \phi y)$, so that, if f is a particular solution, $\phi^{-1}f(\phi x, \phi y)$ is a general one. Babbage generalises to say that $\psi^{n,m}(x, y) = \phi^{-1}f^{n,m}(\phi x, \phi y)$ and that if more than two variables are present the substitution to make is

$$
\psi(x_1, x_2, \ldots x_i) = \phi^{-1}f^{1,1} \ldots (\phi x_1, \phi x_2, \ldots \phi x_i)
$$

and

$$
\psi^{n,m,p,\ldots}(x_1, \ldots x_i) = \phi^{-1}f^{n,m,p,\ldots}(\phi x_1, \ldots \phi x_i).
$$

There is one difficulty, for returning to the simplest case of these generalisations, Babbage says: 'After repeated endeavours I have been unable to find any particular case which will satisfy the equation $\psi^{2,1}(x, y) = \psi^{1,2}(x, y)$.' (*Ibid.*, p. 217.)

The next two pages attempt to show that finding a particular solution is impossible in this case. He concludes:

> This train of reasoning I offer with considerable hesitation, well aware of the extreme difficulty of reasoning correctly on a subject so very general, and which, from its novelty, the mind has not been sufficiently habituated to consider, so as to rely with confidence on any lengthened process of reasoning. I thought it, however, right to mention this proof, that those who may seek for particular cases, might first enquire whether the equation be possible. (*Ibid.*, p. 219.)

Before turning to differential functional equations, there is a digression to a consideration of first order functional equations. He examines a new method to solve equations like $F(x, \psi x, \psi \alpha x, \ldots \psi \alpha^n x) = 0$, where $\alpha^{n+1} x = x$, which according to Problem VII of Part I is the form that all functional equations of the first order may be reduced to. He explains:

> There is a remarkable difference between the former methods and the present one.
>
> Those which I have already given [,] always led to the general solution, and perhaps in some cases, to the complete one; these, on the contrary, which I shall now propose, always conduct us directly to a particular solution, which does not contain even an arbitrary constant. It has, however, several advantages: it is the most direct method with which we are yet acquainted; and if by any means we could introduce into these solutions an arbitrary constant, it would afford us general ones: this is a step which is wanting to connect it with former methods. (*Ibid.*, p. 229.)

To connect this digression with differential functional equations, he adds: 'In the case of differential functional equations, this step is supplied by the integrations which are necessary, and we thus arrive at their general solutions.' (*Ibid.*, pp. 229–30.)

There is therefore a link in method between the first order functional equations and the differential functional equations. One example will suffice to illustrate the technique used to solve the former case. Problem XXXI on p. 230 is to solve the general equation $F(x, \psi x, \psi \alpha x) = 0$, where $\alpha^2 x = x$.

This is solved by expressing ψx in terms of x and $\psi \alpha x$. Assuming this can be done, $\psi x = F_1(x, \psi \alpha x)$. Replace x by αx, obtaining $\psi \alpha x = F_1(\alpha x, \psi x)$, since $\alpha^2 x = x$, and eliminating $\psi \alpha x$, $\psi x = F_1[x, F_1(\alpha x, \psi x)]$, from which (with luck!) ψx may be determined as a function of x.

He illustrates this by solving $(\psi x)^p \psi(a - x) = x^n$. Solving for ψx, $\psi x = [x^n / \psi(a - x)]^{1/p}$. Replacing x by $a - x$ (the αx here), $\psi(a - x) = [(a - x)^n / \psi(x)]^{1/p}$, Therefore

$$\psi x = \left\{ \frac{x^n}{[(a - x)^n / \psi(x)]^{1/p}} \right\}^{1/p}$$

and hence

$$\psi x = \left[\frac{x}{(a - x)^{1/p}} \right]^{np/(p^2 - 1)}.$$

This method has to be used with caution, for if we attempt to solve equations like $\psi x = (a - x)\psi\alpha x$, where $\alpha^2 x = x$ (*ibid.*, p. 233), then the method gives $\psi x = (a - x)\psi\alpha x$, $\psi\alpha x = (a - \alpha x)\psi\alpha^2 x = (a - \alpha x)\psi x$, and therefore $\psi x = (a - x)(a - \alpha x)\psi x$, which leads nowhere!

Now to look at the very interesting differential functional equations. The first one, Problem XXXIII on p. 235, is to solve $\psi\alpha x = \mathrm{d}\psi x/\mathrm{d}x$ where $\alpha^2 x = x$. Replacing x by αx,

$$\psi\alpha^2 x = \psi x = \frac{\mathrm{d}\psi\alpha x}{\mathrm{d}\alpha x}.$$

Differentiating,

$$\frac{\mathrm{d}\psi x}{\mathrm{d}x} = \frac{\mathrm{d}}{\mathrm{d}x}\left[\frac{\mathrm{d}\psi\alpha x}{\mathrm{d}\alpha x}\right]$$

therefore

$$\psi\alpha x = \frac{\mathrm{d}}{\mathrm{d}x}\left[\frac{\mathrm{d}\psi\alpha x}{\mathrm{d}\alpha x}\right]$$

but also,

$$\frac{\mathrm{d}\psi\alpha x}{\mathrm{d}x} = \frac{\mathrm{d}\psi\alpha x}{\mathrm{d}\alpha x}\frac{\mathrm{d}\alpha x}{\mathrm{d}x},$$

therefore

$$\frac{\mathrm{d}\psi\alpha x}{\mathrm{d}\alpha x} = \frac{\mathrm{d}\psi\alpha x}{\mathrm{d}x}\left(\frac{\mathrm{d}\alpha x}{\mathrm{d}x}\right)^{-1},$$

and hence

$$\psi\alpha x = \frac{\mathrm{d}}{\mathrm{d}x}\left[\frac{\mathrm{d}\psi\alpha x}{\mathrm{d}x}\left(\frac{\mathrm{d}\alpha x}{\mathrm{d}x}\right)^{-1}\right],$$

if

$$z = \psi\alpha x,$$

$$z = \frac{\mathrm{d}}{\mathrm{d}x}\left[\frac{\mathrm{d}z}{\mathrm{d}x}\left(\frac{\mathrm{d}\alpha x}{\mathrm{d}x}\right)^{-1}\right],$$

which is a differential equation in z which can, in theory, be solved.

To illustrate this, on p. 236 the problem $\psi(a-x) = d\psi x/dx$ is considered. Here

$$\frac{d\alpha x}{dx} = \frac{d(a-x)}{dx} = -1.$$

This leads to the differential equation

$$z = \frac{d}{dx}\left(\frac{-dz}{dx}\right),$$

or $d^2z/dx^2 + z = 0$, from which $z = b\cos x + c\sin x$, where b and c are arbitrary constants.

We have then the functional equation for ψ, $\psi(a-x) = b\cos x + c\sin x$. Hence $\psi x = b\cos(a-x) + c\sin(a-x)$. The constants b and c must be adjusted to suit the original equation,

$$\psi(a-x) = \frac{d\psi x}{dx}.$$

This gives $b\cos x + c\sin x = b\sin(a-x) - c\cos(a-x)$. Equating coefficients in $\cos x$ and $\sin x$, since the equation is identical, $b = b\sin a - c\cos a$ and $c = -b\cos a - c\sin a$. (*Ibid.*, p. 236.) And these are consistent if $c = -b\cos a/(1+\sin a)$. I take Babbage's $-b\cos a/(1-\sin a)$ to be a misprint.

This gives a solution to the equation as $\psi x = b\cos(a-x) - b\cos a/(1-\sin a)\sin(a-x)$, where b is an arbitrary constant.

It will be seen that the method consists of finding a particular solution and making it as general as possible by the introduction of an arbitrary constant. This constant arises naturally in problems of this type where an indefinite integration has to be performed somewhere. It is interesting to note that a functional equation involving first derivatives leads to an auxiliary differential equation of the second order. The solution of this differential equation gives two arbitrary constants which on substitution in the original equation turn out to be dependent. It is clear that the solution given is not nearly general enough, containing none of the arbitrary functions normally associated with first functional equations of a less complex nature, but this type of functional equation is very difficult and Babbage has done well to find a solution with any sort of generality.

He next tries the complex process of generalisation. To solve

$$\psi \alpha x = \frac{\mathrm{d}^n \psi x}{\mathrm{d}x^n}$$

where $\alpha^p x = x$ (Problem XXXIV, *ibid.*, pp. 237–8), he replaces x in turn by $\alpha x,\ \alpha^2 x,\ \ldots \alpha^{p-1}x$. This gives

$$\psi \alpha x = \frac{\mathrm{d}^n \psi x}{\mathrm{d}x^n},$$

$$\psi \alpha^2 x = \frac{\mathrm{d}^n \psi \alpha x}{(\mathrm{d}\alpha x)^n},$$

$$\vdots$$

$$\psi \alpha^{p-1} x = \frac{\mathrm{d}^n \psi \alpha^{p-2} x}{(\mathrm{d}\alpha^{p-2} x)^n},$$

$$\psi \alpha^p x = \frac{\mathrm{d}^n \psi \alpha^{p-1} x}{(\mathrm{d}\alpha^{p-1} x)^n} = \psi x,\ \text{since } \alpha^p x = x.$$

Combining these equations,

$$\psi \alpha^p x = \psi x = \frac{\mathrm{d}^n}{(\mathrm{d}\alpha^{p-1} x)^n}\ \frac{\mathrm{d}^n}{(\mathrm{d}\alpha^{p-2} x)^n} \cdots \frac{\mathrm{d}^n \psi x}{(\mathrm{d}x)^n},$$

which on resolution is a differential equation of the $p n$th order.

This result leads to two further generalisations. On p. 238, he suggests that

$$\mathrm{F}\left(x,\ \psi x,\ \psi \alpha x,\ \frac{\mathrm{d}\psi x}{\mathrm{d}x}\right) = 0$$

where $\alpha^p x = x$, can be solved, providing that $\psi \alpha x$ can be expressed in terms of the other variables. If this can be done and

$$\psi \alpha x = \mathrm{F}_1\left(x,\ \psi x,\ \frac{\mathrm{d}\psi x}{\mathrm{d}x}\right),$$

then, as in the previous problem, x is replaced in turn by αx, $\alpha^2 x,\ \ldots \alpha^{p-1}x$. On combination these lead to a complicated but recognisable subsidiary differential equation.

He asserts on p. 240, that exactly the same method can be used to solve

$$\mathrm{F}\left(x,\ \psi x,\ \psi \alpha x,\ \ldots \psi \alpha^{p-1} x,\ \frac{\mathrm{d}^n \psi \alpha x}{\mathrm{d}x^n},\ \frac{\mathrm{d}^m \psi \alpha x}{\mathrm{d}x^m},\ \ldots\right) = 0,$$

where $\alpha^p x = x$.

The whole subject is getting very complex. On p. 241, he says:

> Other more complicated equations containing the various orders of functions, and their differentials may be reduced to those of the first order by the same means, but great difficulties still remain; it is by no means easy to discover particular solutions of the original equations, and even when these are found, the functional equations of the first order which remain to be solved, are of considerable difficulty. I shall therefore refrain from giving any more examples, and proceed to show how functional equations involving definite integrals may be reduced to those we have already treated.

He only gives one example involving integral equations (Problem XXXVII, *ibid.*, p. 242) and this one is not altogether satisfying. To solve, as he puts it, $\int dx, \psi^2 x = \psi a$, where the integration is performed between the limits $x = 0$ and $x = a$. Suppose there is a function of two variables $\phi(x, v)$ such that $\phi(x, v) - \phi(0, v) = v$. He asserts that this is a functional equation that can be solved.

Suppose also that the indefinite integral

$$\int dx \psi^2 x = \phi(x, \psi a).$$

This is consistent, for then

$$\int_0^a dx \cdot \psi^2 x = \phi(a, \psi a) - \phi(0, \psi a) = \psi a.$$

Then, noting that differentiation and indefinite integration are inverse processes,

$$\psi^2 x = \frac{d}{dx}[\phi(x, \psi a)],$$

which is a differential functional equation that can be solved.

The method seems artificial, but does suggest a way in which such problems might be tackled.

After this not too successful sortie into integral equations, he ends the paper by a consideration of differential functional equations involving two variables.

The easiest one is Problem XXXVIII (*ibid.*, p. 245), which is to solve

$$\psi(x, y) = \frac{d\psi(x, \alpha y)}{dx}$$

where $\alpha^2 y = y$. Replace y by αy: then

$$\psi(x, \alpha y) = \frac{\mathrm{d}\psi(x, y)}{\mathrm{d}x}.$$

Differentiating, we obtain

$$\frac{\mathrm{d}\psi(x, \alpha y)}{\mathrm{d}x} = \psi(x, y) = \frac{\mathrm{d}^2\psi(x, y)}{\mathrm{d}x^2};$$

it being assumed, of course, that the derivatives are partial with respect to x. This gives a simple differential equation

$$\psi(x, y) = \frac{\mathrm{d}^2\psi(x, y)}{\mathrm{d}x^2},$$

which has solution $\psi(x, y) = \mathrm{E}^x \phi y + \mathrm{E}^{-x} \phi_1 y$, where ϕ and ϕ_1 are arbitrary functions. It should be noted that Babbage used the symbol E where we would write e.

Substituting this solution in the original equation,

$$\mathrm{E}^x \phi y + \mathrm{E}^{-x} \phi_1 y = \frac{\mathrm{d}\psi(x, \alpha y)}{\mathrm{d}x} = \frac{\mathrm{d}}{\mathrm{d}x}(\mathrm{E}^x \phi \alpha y + \mathrm{E}^{-x} \phi_1 \alpha y)$$
$$= \mathrm{E}^x \phi \alpha y - \mathrm{E}^{-x} \phi_1 \alpha y.$$

This means that $\phi y = \phi \alpha y$ and $\phi_1 y = -\phi_1 \alpha y$ which are both soluble functional equations.

An attempt to solve

$$\psi(x, y) = \frac{\mathrm{d}^n \psi(x, y)}{\mathrm{d}x^n},$$

where $\alpha^p y = y$ (Problem XXXIX, *ibid.*, p. 246), leads to the subsidiary differential equation

$$\psi(x, y) = \frac{\mathrm{d}^{np} \psi(x, y)}{\mathrm{d}x^{np}},$$

which can always be solved if np is a whole number. This leads Babbage to a speculation. Suppose we take $n = \frac{1}{2}$ and $p = 2$. Then $np = 1$, $\alpha^2 y = y$ and the equation to be solved is

$$\psi(x, y) = \frac{\mathrm{d}^{1/2} \psi(x, \alpha y)}{\mathrm{d}x^{1/2}}.$$

Now since $np = 1$, this has a solution from

$$\psi(x, y) = \frac{\mathrm{d}\psi(x, y)}{\mathrm{d}x}$$

of $\psi(x, y) = E^x \phi y$ where ϕ is arbitrary. His imagination having been roused, he continues, 'Not only may the index of differentiation become fractional, but the index of the order of a function may be a fraction or even a variable quantity, and such equations as the following might occur

$$\frac{\mathrm{d}^{1/2}\psi^{1/2}x}{\sqrt{\mathrm{d}x}} = \frac{\mathrm{d}^{1/3}\psi^n x}{\mathrm{d}n^{1/3}},$$

(*ibid.*, pp. 248–9). The 'n' in the denominator is obviously a misprint for 'x'.

He goes on to play about with such functions. All that need be said is that provided such operators are properly defined, it is valid to use them. Babbage was no doubt passing from derivatives and functions of positive integral order to those of a fractional order in much the same way that indices are treated, appealing most probably to a form of the 'principle of the permanence of equivalent forms' which he was a few years later to enunciate in his unpublished 'Philosophy of Analysis'.

The summary and hopes for the future are found on the last page of 'An essay towards the calculus of Functions, Part II':

> To complete the outline of this new method of calculation, it would be necessary to treat of equations involving two or more functional characteristics, and to explain methods of eliminating all but one of them; these lead to a variety of interesting and difficult enquiries, and will probably be of considerable use in completing the solutions of partial differential equations; it would also be proper to consider the maxima and minima of functions, and to apply to this subject the method of variations; these are points of considerable difficulty, and although I have made some little progress in each of them, I shall forbear for the present any further discussion on this subject.

Babbage concludes this paper:

> The doctrine of functions is of so general a nature, that it is applicable to every part of mathematical enquiry, and seems eminently qualified to reduce into one regular and uniform system the diversified methods and scattered artifices of the

> modern analysis; from its comprehensive nature, it is fitted for
> the systematic arrangement of the science, and from the new
> and singular relations which it expresses, it is admirably
> adapted for further improvements and discoveries.

This unfortunately proved to be a prediction not fulfilled by
Babbage, or it seems, any other mathematician.

The remainder of Babbage's work on the calculus of functions
is disappointing in the extreme. He never succeeded in following
up the aims of the last paragraph of the second essay, and most of
the subsequent work represents a regression of ideas which we
will describe as briefly as possible.

In chronological order, the next paper was published in the
Philosophical Transactions, 1817, **107**, 197–216. It was read on 17
April 1817 and has the title: 'Observations on the analogy which
subsists between the calculus of functions and other branches of
analysis'.

The major purpose of this paper was for the writer to demonstrate some of his thought processes and inspiration for his
creativity in the calculus of functions. The paper adds little to the
mathematical content of the subject, but it is always interesting to
hear Babbage think aloud and demonstrate the source of his
ideas.

He begins (*ibid.*, p. 197):

> The employment of such an instrument may, perhaps, create
> surprise in those who have been accustomed to view this science
> as one which is founded on the most perfect demonstration,
> and it may be imagined that the vagueness and errors which
> *analogy*, when unskilfully employed, has sometimes introduced
> into other sciences, would be transferred to this.
>
> It is, however, only as a guide to point out the road to
> discovery, that analogy should be used, and for this purpose it is
> admirably adapted.

It is to be understood then, that the contents of the paper are to be
conceived in an heuristic, rather than a rigorous, manner.

He continues:

> It is usually more difficult to discover than to demonstrate any
> proposition; for the latter process we may have rules, but for the
> former we have none. The traces of those ideas which, in the
> mind of the discoverer of any new truth, connect the unknown
> with the known, are so faint, and his attention is so much more

intensely directed to the object, than to the means by which he attains it, that it not unfrequently happens, that while we admire the happiness of a discovery, we are totally at a loss to conceive the steps by which its author ascended to it.

From these considerations, I think it will appear that any successful attempt to embody into language those fleeting laws by which the genius of the inventor is insensibly guided in the exercise of the most splendid privilege of intellect, would contribute more to the future progress of mathematical science than any thing that has hitherto been accomplished. Amidst the total absence of all attempts of this kind, the following illustrations of one of the most obvious assistants of the inventive faculty, will not, I hope, be considered useless. (*Ibid.*, pp. 197–8.)

I believe that the historian of mathematics has a role to play in the development of the subject, complementary to that of the textbook writer. The latter presents his topic in 'tidy' form with the principles enunciated in logical order, a sequence which is usually always different to that of discovery. The historian on the other hand presents the same material through a succession of guesses, logic, hunches and mistakes, and it is often through a study of error that we arrive at a much better appreciation of the subject and can reach the heart of the matter more deeply than by means of a formal approach. A modern textbook on calculus, for example, would usually begin with a discussion of the concept of limit, followed by continuity and differentiability. A book on the history of calculus would describe, by means of the acceptance and rejection of such methods as exhaustion, fluxions, differentials, prime and ultimate ratios, moments, nascent increments and vanishing quantities, why the modern concept of limit is important and has to be so carefully defined.

It is, therefore, very valuable when a mathematician like Babbage says something about the process of discovery as opposed to the process of justification and describes, in his own case, how mathematical creativity works. Having said this, it is necessary to add the rider that it was one of Babbage's characteristic weaknesses, demonstrated in such things as the calculus of functions and the analytical engine, not to succeed in bringing to completion what he said he intended to do. We will find in this paper some indication of his processes, but not many, and for a further account, refer to chapter 5, 'The Philosophy of Analysis'.

His first analogy is to derive the solution of certain problems in the calculus of functions by a method related to the calculation of algebraic indeterminate forms. The link, he says, is that algebra is concerned with variable quantities capable of all degrees of *magnitude*, while the calculus of functions deals with arbitrary characteristics capable of all varieties of *form*. He makes the statement: 'Such is the fraction $(a^x - b^x)/x$ which when $x = 0$ becomes $(1-1)/0 = 0$, and yet its real value is well known to be $\log a/b$,' (*Ibid.*, p. 198.) In view of the heuristic nature of the paper, we can accept such uncritical evaluation of limits, but when this idea is transferred to functions, there is worse to come. Considering the expression $[\{\psi x - \psi(1/x)\}/\psi x]$, where ψx and $\psi(1/x)$ are both zero (as, for example, if $\psi x = v\phi x + v^2\phi_1 x + \ldots$), so that $\psi(1/x) = v\phi(1/x) + v^2\phi_1(1/x) + \ldots$ and if $v = 0$ then both $\psi x = 0$ and $\psi(1/x) = 0$, he substitutes for the fraction involving

$$\frac{\psi x - \psi(1/x)}{\psi x} = \frac{(v\phi x + v^2\phi_1 x + \cdots) - [v\phi(1/x) + v^2\phi_1(1/x) + \cdots]}{v\phi x + v^2\phi_1 x}$$

which on division by v, and putting $v = 0$ (a contradiction, incidentally), he arrives at the 'vanishing' form of $[\{\psi x - \psi(1/x)\}/\psi x]$ as $[\{\phi x - \phi(1/x)\}/\phi x]$ where ϕ is an arbitrary function! It is difficult to know what this means, if anything.

Next, if $(f_1 x - fx f_1\alpha x)/(1 - fx f\alpha x) = 0/0$ in the case $fx f\alpha x = 1$ and $f_1 x - fx f_1\alpha x = 0$, where $\alpha^2 x = x$, by putting $f_1 x = f_1 x + v\phi_1 x$ and $fx = fx + v\phi x$, he obtains on substitution, and putting $v = 0$,

$$\frac{f_1 x - fx f_1\alpha x}{1 - fx f\alpha x} = \frac{\phi_1 x - fx\phi_1\alpha x - f_1\alpha x\phi x}{fx\phi\alpha x + f\alpha x\phi x},$$

where ϕ is arbitrary. This is useful for solving functional equations of the form $\psi x + fx\psi\alpha x = f_1 x$ (*ibid.*, p. 202) in certain extreme cases. A particular solution, worked out previously, is $\psi x = (f_1 x - fx f_1\alpha x)/(1 - fx f\alpha x)$, which becomes critical when f is of the form $fx f\alpha x = 1$. This can frequently occur, as he shows previously on p. 201, where the general solution of such a functional equation is given as $fx = [\chi(\bar{x}, \alpha\bar{x})]^{x - \alpha x}$, $\chi(\bar{x}, \alpha\bar{x})$ being any function symmetrical in x and αx.

The difficulty is overcome by substituting for the awkward expression in f and f_1 its 'limit', $(\phi_1 x - fx\phi_1\alpha x - f_1\alpha x\phi x)/(fx\phi\alpha x + f\alpha x\phi x)$, which becomes the general solution.

As an example, he cites the case where $\alpha x = -x$ and $fx = f_1 x = 1$ so that the functional equation becomes $\psi x + \psi(-x) = 1$. The general solution will then be $\psi x = [\phi_1 x - \phi_1(-x) + \phi x]/$

$[\phi(-x)+\phi(x)]$, where ϕ and ϕ_1 are both arbitrary functions. It is extraordinary that the solution is correct, and Babbage emphasises his conclusion in triumph: '*Whenever the method of elimination apparently fails, the real value of the vanishing fraction will give the general solution of the equation.*' (*Ibid.*, p. 202.)

Actually, there is an extraordinary cancellation of errors here. If Babbage had applied the limit formula correctly, he would have obtained not $[\phi_1 x - \phi_1(-x) + \phi x]/[\phi(-x)+\phi(x)]$ but $[\phi_1 x - \phi_1(-x) - \phi x]/[\phi(-x)+\phi(x)]$, which on substitution leads to $\psi(x) + \psi(-x) = -1$, which is not the equation to be solved. However, this is compensated by an earlier error. A minus sign has been left out of the limit formula which correctly written should be $(\phi x f_1 \alpha x + f x \phi_1 \alpha x - \phi_1 x)/(f x \phi \alpha x + f \alpha x \phi x)$, which leads, on substitution $f x = f_1 x = 1$, $\alpha x = -x$, to $[\phi(x)+\phi_1(-x)-\phi_1(x)]/[\phi(x)+\phi(-x)]$. This does, in fact, satisfy $\psi(x)+\psi(-x)=1$.

The theory, although expressed rather bluntly here, is probably sound if reformulated carefully in terms of limits. To give another illustrative example, we might take

$$\alpha x = \frac{1}{x}, f(x) = x \quad \text{and} \quad f_1 x = \sqrt{x}$$

so that

$$\frac{f_1 x - f x f_1 \alpha x}{1 - f x f \alpha x} = \frac{\sqrt{x} - x(1/\sqrt{x})}{1 - x(1/x)}.$$

The functional equation to solve then becomes $\psi(x)+x\psi(1/x)=\sqrt{x}$, which then has solution

$$\psi x = \frac{(\phi x/\sqrt{x})+x\phi_1(1/x)-\phi_1(x)}{(1/x)\phi(x)+x\phi(1/x)},$$

which on substitution is correct.

The next result is one of elimination. Suppose we are given a functional equation of the form

$$F(\psi x, \psi \alpha x, \ldots \psi \alpha^{n-1} x, \bar{x}, \overline{\alpha x}, \ldots \overline{\alpha^{n-1} x}) = 0,$$

where $\alpha^n x = x$, the bars indicating symmetry in the variables concerned. Then if x is replaced in turn by $\alpha x, \ldots \alpha^{n-1} x$, the equations

$$F(\psi x, \psi \alpha x, \ldots \psi \alpha^{n-1} x, \bar{x}, \overline{\alpha x}, \ldots \overline{\alpha^{n-1} x}) = 0,$$

$$F(\psi \alpha x, \psi \alpha^2 x, \ldots \psi x, \bar{x}, \overline{\alpha x}, \ldots \overline{\alpha^{n-1} x}) = 0,$$

and

$$F(\psi\alpha^{n-1}x, \psi x, \ldots \psi\alpha^{n-2}x, \bar{x}, \overline{\alpha x}, \ldots \overline{\alpha^{n-1}x}) = 0$$

are obtained. These equations considered together are symmetrical in ψx, $\psi\alpha x$, ... so that whichever is eliminated, the same result will be obtained. This means that $\psi x = \psi\alpha x = \cdots = \psi\alpha^{n-1}x$, and the equation reduces to $F(\psi x, \psi x, \ldots \psi x, \bar{x}, \overline{\alpha x}, \ldots \overline{\alpha^{n-1}x}) = 0$ from which ψx may be obtained (*ibid.*, p. 204).

Babbage next remarks on a similarity between the roots of unity to solutions of the equation $\psi^n x = x$ (*ibid.*, p. 205). In the equation $r^n = 1$, if r_1 is a root, then any power of r_1 is also a root. Further, if n is a prime number and r_1 any root except unity, then r_1, $r_1^2, \ldots r_1^{n-1}$ are all distinct roots. In a similar manner, if αx is a solution of $\psi^n x = x$, so also is $\alpha^m x$, and, if n is prime, αx, $\alpha^2 x, \ldots \alpha^{n-1}x$ are all different solutions of the equation. Also, for example, any solution of $\psi^3 x = x$ satisfies $\psi^6 x = x$. Babbage was clearly fascinated by the relation of the roots of unity and constantly refers to them in many branches of his work.

He ends this short paper with two analogies between the calculus of functions and properties of differential equations. The first (*ibid.*, p. 208) is a comparison of the differential equation

$$0 = y \, dx^n + A_x \, dx^{n-1} \, dy + B_x \, dx^{n-2} \, d^2y + \cdots + N_x \, d^ny + X \, dx^n$$

with the functional equation

$$0 = \psi x + A_x\psi\alpha x + B_x\psi\alpha^2 x + \cdots + N_x\psi\alpha^{n-1}x + X.$$

The similarity is that in each case the last term X can be eliminated by finding a particular solution. Then the substitution $y = f(x) + z$ or $\psi(x) = f(x) + \phi(x)$, where $f(x)$ is the particular solution in either case, leads to a differential or functional equation with the term in X eliminated.

The second analogy (*ibid.*, p. 213) is to solve a functional equation by a method used in first order differential equations of multiplying by a particular integral to make one side of the equation integrable. In differential equations, to solve $(dy/dx) + Py = Q$, where P and Q are functions of x, multiplying by F, $F(dy/dx) + FPy = FQ$. Then if F is chosen so that $FP = dF/dx$, the left-hand side becomes $F(dy/dx) + (dF/dx)y$ which is $d/dx(Fy)$. To find such an F, solve the simple differential equation $FP = dF/dx$, or $P \, dx = dF/F = \log_e F$. Thus $F = e^{\int P \, dx}$.

A similar technique may be used to solve the functional equation $\psi x + fx\psi\alpha x = f_1 x$ where $\alpha^2 x = x$. Multiplication of both sides by any function ϕx gives $\phi x\psi x + \phi x\, fx\psi\alpha x = \phi x f_1 x$. If ϕx is now chosen so that $\phi x = f\alpha x\phi\alpha x$, the left-hand side becomes symmetrical with respect to ψx and $\psi\alpha x$. This equation, $\phi x = f\alpha x\phi\alpha x$, which may also be expressed as $\phi\alpha x = fx\phi x$, can be solved when fx is given, and the original functional equation completed.

After these two interesting analogies, Babbage admits:

> When I had ascertained the remarkable similitude which exists between the method of functions and the integral calculus, I referred to a treatise on that subject with the express purpose of endeavouring to transfer the methods and artifices employed in the latter calculus, to the cultivation and improvement of the former. (*Ibid.*, pp. 215–16.)

The remarks contained in this paper reveal an interesting connection in Babbage's mind between the different branches of mathematics, but tell us little about the psychology of discovery. At any rate, it shows that the author was skilful in using ideas from simpler results in other fields to suggest more difficult ones.

Before leaving this paper, there is an interesting remark on the calculus of functions made on p. 210. Babbage mentions that a functional equation can have three types of solution. They are:

> 1st. The complete solution, this contains as many arbitrary functions as the nature of the given equation will admit.
> 2d. The particular case, this contains all solutions which are less general than the complete solution, and which are only particular cases of it. If they contain arbitrary functions, I have called them, for the sake of convenience, general solutions.
> 3d. The particular solution, this is a solution which satisfies the equation, and may or may not contain arbitrary functions; its peculiar property is, that it is found from one part only of the equation and independent of the rest.

He adds on p. 211 that these species: 'Bear a strong resemblance to those of differential equations.'

It might be added that all the solutions given by Babbage involve only the second and third types. He has no way of obtaining the complete solution of the first type to any functional equation.

The next paper, 'Solutions of some problems by means of the calculus of functions', was published in the *Journal of Science and the Arts*, London, 1817, **2**, 371–9. This journal, which was edited at the Royal Institution of Great Britian, seems to have been used by Babbage for his papers of lesser importance, possibly with a view to reaching the layman rather than the expert.

This particular paper is mainly concerned with geometrical applications of the calculus of functions, and as this idea is dealt with more fully in a paper written a year later, it will suffice to give just one example. This is Problem I, taken from pp. 371–3, in which the question is to find a curve ABC in which, if the rectangle containing any two abscissae, AD and AE, is equal to a given square, the rectangle containing the two corresponding ordinates BD and CE, will be equal to another given square.

In terms of the calculus of functions, this means that if the equation of the curve is $y = \psi x$, and the given squares are a^2 and c^2 respectively, then if $AD = x$, $AE = (a^2/x)$, $BD = \psi x$, $CE = \psi(a^2/x)$ and the functional equation becomes $\psi x \cdot \psi(a^2/x) = c^2$.

The technique used is to look first for a particular solution containing an arbitrary constant. This is done by assuming that $\psi x = Ax^n$. On substitution, $\psi(a^2/x) = A(a^2/x)^n$, so that the equation becomes $Ax^n \cdot A(a^2/x)^n = c^2$, or $A^2 a^{2n} = c^2$ so that $A = ca^{-n}$. This yields the particular solution

$$\psi x = c\left(\frac{x}{a}\right)^n,$$

so that n is the required arbitrary constant. He now refers to Problem VIII of 'An essay towards the calculus of functions, Part I' (*Philosophical Transactions*, 1815, **105**, 406). Here it is proved that in place of any arbitrary constant in a solution to a functional equation, may be substituted an arbitrary function which is symmetrical in the terms involving ψ in the original equation. In this case, it means that n may be replaced by any function symmetrical in ψx and $\psi(a^2/x)$. The general solution therefore becomes

$$\psi x = c\left(\frac{x}{a}\right)^{\chi(\bar{x},\bar{a}^2/x)}$$

It is easy to see that this satisfies $\psi x \cdot \psi(a^2/x) = c^2$ for any such function χ. He remarks further that since n could be any sort of number, positive, negative or fractional, then all parabolae and hyperbolae of any degree, certainly have this property.

An even shorter paper 'Note respecting elimination' is included in the next volume of the *Journal of Science and the Arts*, London (1817, **3**, 355–7).

The mathematical content is hardly worth mentioning. It concerns finding a root quickly from a series of equations symmetrical in the variables concerned, i.e. '$F(x, y, z, v, \&c.) = 0$ $F(y, z, v, \&c.) = 0$, &c.' (*Ibid.*, p. 355.)

The method is simply that if, for example, $F(x, y) = 0$ and $F(y, x) = 0$, then, putting $x = y$, the equation $F(x, x) = 0$ will give a solution. In the example supplied (*ibid.*, p. 356), if $x^2 + ay = b$ and $y^2 + ax = b$, then one solution is immediately obtained from $x^2 + ax = b$, namely $x = y = -(a/2) + \sqrt{[b + (a^2/4)]}$.

Similarly, if $F[x, f(x, y)] = 0$ and $F[f(x, y), x] = 0$, then put $v = f(x, y)$ so that $F(x, v) = 0$ and $F(v, x) = 0$. A solution is then obtained $v = x$, $F(x, x) = 0$ and $f(x, y) = x$.

Babbage wrote a further paper relevant to this work, 'On the application of analysis to the discovery of local theorems and porisms' (*Transactions of the Royal Society of Edinburgh*, 1823, 337–52), and a short booklet, *Examples of the Solutions of Functional Equations*, Cambridge, 1820.

The first of these is concerned mainly with applications of the calculus of functions to geometry and I will defer discussion of this paper until chapter 6 on Babbage's geometrical work. The second was intended as part of the final salvo of the Analytical Society towards introducing Continental mathematics to this country. Babbage had previously co-operated with J. F. W. Herschel to produce *Memoirs of the Analytical Society*, Cambridge, 1813, and followed, in conjunction also with George Peacock, with the translation of S. F. Lacroix's work, *An Elementary Treatise on the Differential and Integral Calculus*, translated from the French with an Appendix and Notes, Cambridge, 1816. We have seen how it was their intention to produce a book of examples in modern notation, and the result was duly published in 1820. Peacock supplied the worked examples in differential and integral calculus, Herschel those on equations of differences, equations of mixed differences and interpolation, and Babbage supplied the section on the calculus of functions.

Babbage's contribution consisted of eighty-three worked examples spread over forty-two pages. Most of these were taken from his two-part essay in the *Philosophical Transactions* of 1814 and 1815, but there is some new material. He begins the *Examples of the Solutions of Functional Equations* (p. 1):

> The object of the following Examples of Functional Equations, is to render a subject of considerable interest, more accessible to

mathematical students, than it has hitherto been. It is, perhaps, that subject of all others, which most requires the assistance of particular instances, in order fully to comprehend the meaning of its symbols, which are of the most extreme generality; that assistance is also more particularly required in this branch of science, in consequence of its never yet having found its way into an Elementary Treatise.

After an explanation of the notation and the idea of periodic functions (in the involutionary sense) he considers the equation $\psi^2 x = x$ and $\psi^n x = x$ by methods similar to those in the 'Essay'. In the course of solving $\psi^n x = x$ he refers to Horner's general solution of this equation as

$$\psi x = \phi^{-1} \, \frac{a + b\phi(x)}{c - \left(\dfrac{b^2 - 2bc \, \cos \, (2k\pi/n) + c^2}{(2 + 2 \cos \, (2k\pi/ln))a} \right)\phi(x)}$$

where $\phi(x)$ is an arbitrary function and a, b, and c are arbitrary constants.[4]

The method for solving $F(x, \psi x, \psi \alpha x) = 0$, where $\alpha^2 x = x$ by replacing x by αx as given in the 'Essay', is then considered, and Babbage extends the work here by giving examples of techniques to use when the general method does not work. To solve, for example, $\psi x = \psi(1/x)$ (*ibid.*, p. 10) if the method of replacing x by $(1/x)$ is used, we only obtain the same equation, $\psi(1/x) = \psi x$.

He suggests instead that we start by considering the more general equation $\psi x = a\psi(1/x) + v\phi x$ where a, v and ϕx are all arbitrary. Replacing x by $1/x$ gives $\psi(1/x) = a\psi x + v\phi(1/x)$. $\psi(1/x)$ can now be eliminated and we are left with

$$\psi x = a \left[a\psi x + v\phi(1/x) \right] + v\phi x \quad \text{or} \quad \psi x = \left[\frac{a\phi(1/x) + \phi x}{1 - a^2} \right] v.$$

It follows that the solution of $\psi x = \psi(1/x)$ will be

$$\psi x = a \overset{Lt}{\to} 1, \ v \to 0 \left[\frac{v}{1 - a^2} \right] \left[\phi \left(\frac{1}{x} \right) + \phi x \right].$$

Assuming this double limit can be calculated, and Babbage achieves this by the dubious method of putting $a = 1 + t$, $v = t$ so that as $t \to 0$ the limit becomes $-\frac{1}{2}$, the general solution may be left as $\psi x = \phi(1/x) + \phi x$ where ϕ is an arbitrary function. He generalised this to say that a better solution is $\psi(x) = \chi(\bar{x}, \bar{1}/x)$, i.e. any function symmetrical in x and $1/x$, and that the general solution of any functional equation $\psi x = \psi \alpha x$, where $\alpha^2 x = x$, is $\psi x = \chi(\bar{x}, \overline{\alpha x})$.

When such an uncritical method has been used it is aggravating to admit that the solution is undoubtedly correct. One can, however, give vent to feeling by saying that the general solution $\psi x = \chi(\bar{x}, \overline{\alpha x})$ is obvious from the equation $\psi x = \psi \alpha x$ immediately, without having to go through this extraordinary calculation.

It is interesting to note one example arising from this method (*ibid.*, p. 14) in view of its relation to Laplace's work, and I quote Babbage's reasoning in full.

> Given the equation
>
> $$\{\psi x\}^2 + \left\{\psi\left(\frac{\pi}{2} - x\right)\right\}^2 = 1.$$
>
> This is the equation on which the composition of forces is made to depend in the Mécanique Céleste, Paris, 1797, Tome 1, p. 5.
> Put $\psi_1 x$ for $(\psi x)^2$ then it becomes
>
> $$\psi_1 x + \psi_1\left(\frac{\pi}{2} - x\right) = 1$$
>
> which is a particular case of
>
> $$\psi_1 x + a\psi_1\left(\frac{\pi}{2} - x\right) = 1 + v\phi x.$$
>
> Substituting $\pi/2 - x$ for x, this gives
>
> $$\psi_1\left(\frac{\pi}{2} - x\right) + a\psi_1 x = 1 + v\phi\left(\frac{\pi}{2} - x\right),$$
>
> and eliminating
>
> $$\psi_1\left(\frac{\pi}{2} - x\right),$$
>
> we have
>
> $$\psi_1 x = \frac{1}{1+a} - \frac{\phi x - a\phi\left(\frac{\pi}{2} - x\right)v}{1 - a^2}$$
>
> making $a = 1$ and $v = 0$, and changing ϕx into $2\phi x$, we have
>
> $$\psi_1 x = \frac{1}{2} - \phi x + \phi\left(\frac{\pi}{2} - x\right)$$

and therefore

$$\psi x = \sqrt{\frac{1}{2} - \phi x + \phi\left(\frac{\pi}{2} - x\right)}$$

The solution is again correct. There is no need to worry that

$$\begin{array}{c} \text{Lt} \\ a \to 1 \\ v \to 0 \end{array} \frac{v}{1-a^2}$$

has again been taken as $-\frac{1}{2}$. Assuming such a limit does exist, putting

$$\begin{array}{c} \text{Lt} \\ a \to 1 \\ v \to 0 \end{array} \frac{v}{1-a^2} = t,$$

the same result will follow.

The novel method for solving a certain type of first order functional equation is given on p. 26. The equation is $1 + fx(\psi x + \psi \alpha x) - \psi x \psi \alpha x = 0$, where $\alpha^2 x = x$ and fx is symmetrical with respect to x and αx.

The equation may be rewritten

$$fx = \frac{\psi x \cdot \psi \alpha x - 1}{\psi x + \psi \alpha x}.$$

The novelty of the method is that this is now differentiated with respect to ψx and $\psi \alpha x$, obtaining the results

$$\frac{dfx}{d\psi x} = \frac{(\psi x + \psi \alpha x)\left(\psi \alpha x + \psi x \dfrac{d\psi \alpha x}{d\psi x}\right) - \left(1 + \dfrac{d\psi \alpha x}{d\psi x}\right)(\psi x \psi \alpha x - 1)}{(\psi x + \psi \alpha x)^2}$$

$$= \frac{(\psi \alpha x)^2 + 1 + [(\psi x)^2 + 1]\dfrac{d\psi \alpha x}{d\psi x}}{(\psi x + \psi \alpha x)^2}$$

and similarly

$$\frac{dfx}{d\psi \alpha x} = \frac{(\psi x)^2 + 1 + [(\psi \alpha x)^2 + 1]\dfrac{d\psi x}{d\psi \alpha x}}{(\psi x + \psi \alpha x)^2}.$$

Babbage now appears to take $fx = $ constant, so that the two numerators vanish, giving

$$\frac{\mathrm{d}\psi x}{1+(\psi x)^2}+\frac{\mathrm{d}\psi\alpha x}{1+(\psi\alpha x)^2}=0.$$

This is easily integrated to $\tan^{-1}\psi x + \tan^{-1}\psi\alpha x = C$. But from the initial equation $(\psi x + \psi\alpha x)/(1-\psi\chi\psi\alpha x) = -1/fx$, so that $\tan^{-1}\psi x + \tan^{-1}\psi\alpha x = \tan^{-1}$ $[(\psi x + \psi\alpha x)/(1-\psi x\psi\alpha x) = \tan^{-1}$ $(-1/fx)$. If ϕx is arbitrary, this has solution $\tan^{-1}\psi x = \phi x/(\phi x + \phi\alpha x)\tan^{-1}(-1/fx)$, so that the general solution becomes

$$\psi x = \tan\left[\frac{\phi x}{\phi x + \phi\alpha x}\tan^{-1}(-1/fx)\right].$$

Other problems solved in this paper are of the type

$$\psi(x, y) = \frac{\mathrm{d}\psi(x, a-y)}{\mathrm{d}x},$$

$$\psi(x, y) = \frac{\mathrm{d}^n\psi(x, \alpha y)}{\mathrm{d}x^n},$$

where $\alpha^n x = x$, $\psi^{\overline{2,2}}(x, y) = a$ and $x\psi^{1,2}(x, y) = y\psi^{2,1}(x, y)$.

My aim now is to consider the extent to which Babbage's claim as the inventor of a new calculus can be established. The inventiveness will lie in the systematism of previously known results, for differential and difference equations had been solved for a whole century, and these problems, as Babbage pointed out, are special cases of functional equations. It is still difficult, however, to find any reference to a solution of a purely functional equation by any previous mathematician, and certainly not in the detail that Babbage worked out. Probably the first such problem recorded was solved by D'Alembert. In his papers 'Recherches sur la courbe que forme une corde tendue mise en vibration', *Historische der Akademie, Berlin*, 1747, 214–19, and 'Addition au Memoire sur la courbe que forme une corde tendue mise en vibration', *ibid.*, 1750, 355–60, a problem on vibrating strings reduces itself to the functional equation $f(x+y)-f(x-y) = g(x)\cdot h(y)$ which D'Alembert solves by transformation to a differential equation.

Euler's main contribution was contained in his well-known work on homogeneous functions published in his *Institutiones*

Calculi Integralis, Petropol, 1768, vol. I, in which the functional equation $F(xz, yz) = F(x, y)$ is solved as $F(x, y) = \phi(y/x)$.

It has already been mentioned that Laplace arrived in effect at a functional equation

$$[\phi(\theta)]^2 + \left[\phi\left(\frac{\pi}{2} - \theta\right)\right]^2 = 1$$

on p. 5 of his *Traité de Mécanique Céleste*, Paris, An. VIII, Tome 1, but he did not solve this problem by the functional methods suggested by Babbage, choosing rather to use a longer solution by means of the differential calculus.

There is a paper by G. Monge in the *Mélanges de Philosophie et de Mathematique de la Société Royale de Turin*, published in the *Miscellanea Taurinensia*, 1770–73, **5**, 16–79, entitled 'Mémoire sur la détermination des fonctions arbitraires dans les intégrales de quelques equations aux différences partielles'. The first problem is one in the calculus of functions. It is: 'Trouver quelle doit être la forme de la fonction ϕ dans l'equation $z = \phi V$, pour qu'en fesant $y = \Delta x$ on ait $z = \psi x$; V étant une quantité quelconque donnée en x & y, & les formes des fonctions Δ & ψ étant connues.' (*Ibid.*, p. 19.)

The problem is to find the function ϕ such that $z = \phi[v(x, y)]$ and that, when $y = \Delta(x)$ is substituted for y, $z = \psi(x)$ is obtained, v, Δ and ψ all being known.

The solution given is to put $y = \Delta(x)$ in $v(x, y)$ obtaining $v[x, \Delta(x)]$ which is put as $v'(x)$. The problem now becomes to solve the functional equation $\psi(x) = \phi v'(x)$. If $v' = u$, where $u = u(x)$ is a particular solution, the inverse function of u may be found so that $x = f(u)$. Substituting this function for x, $\psi f(u) = \phi(u)$ is obtained. Both ψ and f are known, so that ϕ can be determined. The solution then becomes $z = \psi f v$.

This is a problem similar to some solved by Babbage but is the only one of its type in Monge's work, and there is no attempt made to generalise the process or to consider more difficult functional equations.

Babbage's great friend and colleague, J. F. W. Herschel, produced two interesting papers on the subject shortly before Babbage's first publication. The first is the chapter 'On functional equations' from the *Memoirs of the Analytical Society*, Cambridge, 1813, written in conjunction with Babbage. His introductory remarks show the mathematical origins of this calculus:

> The integration of equations of partial differences, having introduced arbitrary functions of the independent variables, before such general integrals could be applied to any particular case, it was necessary to determine these functions, so as to satisfy, not only the general equation, but also the particular conditions of the problem, not expressed by that equation. Hence, arose a calculus, whose object might be stated to be, 'the determination of functions from given conditions'.[5]

He goes on to mention attempts to solve such problems, including the one by Monge already referred to; Lagrange, he states, also solved a functional equation on one occasion by the use of a Taylor's series expansion. The normal method for solving these equations, however, was to reduce the problem to one of finite differences. This can usually be done when the equation is a first order one $F(x, \phi_1, \phi_2, \ldots \phi_n) = 0$, where $\phi_1, \phi_2, \ldots \phi_n$ are all given functions of x, and Herschel worked out a few such problems, but the method could not normally be applied to equations of higher functional order. This type of problem created so many complexities that he ended the chapter by asserting: 'The denouement of these difficulties seems reserved for a far more advanced state of Analytical Science, than we can at present boast of having attained.'[6]

Herschel's second contribution contained in a paper entitled 'Consideration of various points of analysis', published in *Philosophical Transactions*, 1814, **104**, 440–68, contains the significant remark: 'Functional Equations have occupied the attention of the most eminent Analysts, and it must be confessed, not without considerable success. Their researches, however, have hitherto extended no further than to such conditions as involve only the unknown function, ϕ without any of its superior or inferior orders, ϕ^2, ϕ^3, ... &c., ϕ^{-1}, &c.[7]

He then shows how the equation $\phi^2 x = x$ can be solved by the method of finite differences.

We have already seen from Bromhead's letter of 9 February 1816, that the calculus of functions was completely new material to English mathematicians. British mathematics was, unfortunately, at a very low ebb at this time, and so this expression of novelty must be supported from foreign sources. Fortunately this can be demonstrated, in the form of a letter dated March 1820 from the great French mathematician, A. L. Cauchy, to Babbage:

> J'ai leu les trois mémoires sur le calcul des fonctions et sur la
> Summation des Séries que vous m'avez fait l'honneur de
> m'adresser. L'importance des sujets que vous y traittez prouve
> que dans la patrie du Newton il existe encore des geometres qui
> travaillent aux progrés de l'analyse. Je me sera inveritable
> plaisir de prendre connaissance de ces memoires, et vous
> remercie beaucoup d'avoir bien voulu m'en faire part. Agréez
> aussi, je vous prie l'assurance de la considération distinguée
> avec laquelle j'ai l'honneur d'être . . .[8]

It can be said with some assurance that no mathematician prior
to Babbage had treated the calculus of functions in such a
systematic way, and problems arising from functional equations
of higher than the first order had been almost totally overlooked.
Babbage must be given full credit as the inventor of a distinct and
important branch of mathematics.

This is borne out by an article by the French mathematician S.
Pincherle, 'Équations et Opérations Fonctionelles', written in the
Encyclopédie des Sciences Mathématiques, Paris and Leipzig, 1912, **5**,
1–78. Speaking historically, he states:

> Dans cet ordre d'idées P. S. Laplace a considéré des équations
> fonctionelles qui se remènant aux équations aux differences
> mêlées; G. Monge a donné de son côté quelques principes
> généraux et quelques procédés de calcul pour ramener aux
> équations aux différences finies certaines classes d'équations
> fonctionelles de formes diverses; Ch. Babbage, après avoir
> traité de nombreux exemples rentrant dans l'une ou l'autre de
> ces classes, s'est occupé, d'une façon générale, des solutions de
> ces équations fonctionelles (qu'on nomme aussi les *intégrales* de
> ces équations) et il distingue ces solutions, en les envisageant au
> point de vue de leur plus ou moins grande indétermination, en
> *solutions générales*, qui sont celles qui contiennent des fonctions
> arbitraires, et en *solutions particulières*, qui sont celles qui ne
> renferment que des constantes arbitraires en nombre fini.[9]

It is good to see such an acknowledgement to Babbage as
virtually the founder of this subject. Elsewhere Pincherle talks
about an equation in his notation $S_\psi^n = 1$, equivalent to $\psi^n x = x$ in
Babbage's, and he actually refers to this as '*l'équation de Babbage*'.[9]

He also adds a result, first discovered in 'An essay towards the
calculus of functions Part 1: 'S*i* ϕ est une solution particulière de
l'équation de Babbage et si f designe une fonction arbitraire,
l'expression $f^{-1}\phi f$ sera la solution générale de l'équation de
Babbage.'[9]

Otherwise there is not much attempt in this paper to develop any of Babbage's results.

It is true that he gave no thought to questions of uniqueness or existence or continuity, but it is perhaps unfair to criticise him on these grounds as the general standards of rigour in mathematics were relatively low at this time, Gauss and Cauchy having barely begun their reformative work.

After Cauchy's letter of 1820 it is interesting to note that in the following year in his *Cours d'analyse de l'École Polytechnique*, Paris, 1821, he actually solved four functional equations considered basic in subsequent theory, namely:

$$f(x+y) = f(x) + f(y),$$

$$f(x+y) = f(x) \cdot f(y),$$

$$f(xy) = f(x) + f(y),$$

and

$$f(xy) = f(x) \cdot f(y).$$

It is actually much easier to solve equations involving a single unknown function with two variables which can conveniently be given arbitrary values than any of Babbage's involving one variable only.

According to J. Aczél, *Vorlesungen über Funktionalgleichungen und ihre Anwendungen*, Basel, 1961, which is an encyclopaedic work on the calculus of functions without, it seems, the author being aware of Babbage's work, functional equations were considered by such mathematicians as Abel, Stokes, Weierstrass, Darboux and Hilbert. While their approach is undoubtedly more rigorous than Babbage's it is doubtful if any of these could rival him in surpassing his contemporaries by the sheer ingenuity of method and generalisation of results.

NOTES

1. C. Babbage, *Passages from the Life of a Philosopher*, London, 1864, p. 435.
2. British Museum Additional Manuscripts 37182, No. 46.
3. *Ibid.*, No. 50.
4. J. Ivory, *Annals of Philosophy*, 1817, **10**, 341–6.
5. *Memoirs of the Analytical Society*, Cambridge 1813, p. 96.
6. *Ibid.*, p. 114.

7. *Philosophical Transactions*, 1814, **104**, 458.
8. British Museum Additional Manuscripts 37182, No. 235.
9. *Encyclopédie des Sciences Mathématiques*, Paris and Leipzig, 1912, **5**, 46.
10. *Ibid.*, 52.

5

'The Philosophy of Analysis'

'The Philosophy of Analysis' is the title of a set of mathematical essays by Charles Babbage, only one of which was ever published. The remainder are bound together and kept in the British Museum Manuscripts Room as Additional Manuscripts 37202.

It is difficult to determine why these essays of generally high quality and much original thought should have been so neglected by Babbage. If published as a book, which would have been his first one, they would no doubt have had a considerable influence on mathematical thought. Round about the year 1830 was one of the most fruitful periods in the history of mathematics, when revolutionary ideas in algebra and geometry were first put forward and the era of modern mathematics could be said to have begun. Two, in particular, of Babbage's papers would have made a major contribution both to the type of algebra generally known as 'modern', to distinguish it from symbolised arithmetic, and to stochastic analysis.

Babbage makes no reference to this proposed work in his *Passages from the Life of a Philosopher*, or in any other of his writings. We can learn of the fate of this work only from a few letters from his friends. E. F. Bromhead, writing on 7 March 1821, is most enthusiastic:

> I am glad you continue to work at the Philosophical Theory of Analysis, I have always considered Notation as the *Grammar* of symbolic language, which can have its false concord, barbarisms, and bad style, as well as any tongue descended from the dispersion of Babel. It will I trust be a Magnum Opus in your hand, and would only want the addition of the Digest to root analysis for ever in this Kingdom.[1]

Unfortunately British mathematics was at a stage when the

number of people, like Babbage's circle of friends, who could understand such abstract reasoning was very limited. I have already mentioned two occasions when Bromhead wrote to Babbage deploring the almost universal ignorance in response to his papers on the calculus of functions. Despite Bromhead's great enthusiasm for his projected work, Babbage was soon to receive a severe setback from his friend Sir David Brewster, the editor of the *Edinburgh Journal of Science*. Brewster's letter, dated 3 July 1821, said:

> I am this moment favoured with your letter of the 20th June, and it is with no inconsiderable degree of reluctance that I decline the offer of any Paper from you. I think, however, you will upon reconsideration of the subject be of opinion that I have no other alternative.
>
> The subjects you purpose for a series of Mathematical and Metaphysical Essays are so very profound, that there is perhaps not a single subscriber to our Journal who could follow them.[2]

This is a somewhat unusual reason for a supposedly learned journal to reject a publication. It is indicative of the intellectual poverty of the times. Babbage might have been sufficiently discouraged to abandon further attempts at publication. This adverse reply does not seem to have strained Babbage's relations with Brewster, for the letter continues by requesting the article 'On notation' for Brewster's *Encyclopaedia*, which Babbage duly produced. Further, in the following year, 1822, the article 'On the theoretical principles of machinery for calculating tables' was printed in Brewster's *Edinburgh Journal*, and five further papers of a non-mathematical nature by Babbage were published in this journal together with an article, 'On porisms', for the *Encyclopaedia*.

However, Babbage did submit his first essay, 'On the influence of signs in mathematical reasoning', to the Cambridge Philosophical Society, and read the paper to them on 16 December 1821. This did not guarantee publication, as a further (undated) letter by his friend Bromhead, in 1822, indicates: 'I see that your Memoir is under perusal at the Cambridge Philosophical Society, it is the only mathematical memoir offered. This is downright affectation, they are attempting things which they do not understand, & neglecting the department in which they might be useful.'[3]

The paper was, however, duly published in 1827. I should hasten to add that the Cambridge Philosophical Society was very slow to start publishing its *Transactions*, the first volume appearing in 1821 and the second not until 1827. The delay did, in fact, give the Society time to consider, and publish in the second volume, another of Babbage's papers, 'On the determination of the general term of a new class of infinite series', read on 3 May 1824.

The most interesting of this series of letters was written by George Peacock, 7 May 1822:

> I shall send your essays tomorrow morning by the coach. I have read the greater part of them over very attentively, a task which you readily acknowledge is of some difficulty considering the manner in which they are written; in some cases I have been completely baffled in my attempt particularly in the latter part of the first essay & in the greatest part of the second. I have before expressed my opinion concerning them; they must form when completed a work of very great interest, abounding as they do with so much of original research & with illustrations of the most interesting kind; the essay on artifices & on questions requiring new methods of analysis will be charming when completed.
> I think the first essay is longer than necessary . . . '[4]

There are several points to be made here. The first is that the reference to the essays on artifices and on questions requiring new methods of analysis, identify the collection referred to in Peacock's letter with 'The Philosophy of Analysis', these being two of Babbage's chapter headings for the latter work. Having said this, it seems unlikely that the unpublished collection in the British Museum is identical with the set of essays read by Peacock, since his remarks about the length and difficulty of the first two are not borne out by an examination of the first two in the collection. The first one, 'On the influence of signs in mathematical reasoning, is not included in the British Museum collection, on account, presumably, of its having been published by the Cambridge Philosophical Society. It is certainly a very long essay, covering over fifty printed pages, but as I remark elsewhere (chapter 8) it is not very profound, and, unless the version Peacock saw was different from the published one, it is difficult to see how a man of his ability could have found it baffling. The second essay, 'On notation', only occupies a couple of pages in the

British Museum manuscript and is again trivial by Babbage's standards, and could therefore not possibly be the same as the one read by Peacock.

Finally, it is most interesting that Peacock admits having read these essays very attentively, for it indicates a possible source of inspiration for his major work, *A Treatise on Algebra*, in 1830. We will consider this in greater detail when examining some very original ideas in Babbage's third essay.

Turning to the British Museum text itself, our first problem will be to relate the contents to Babbage's intention in writing this series of essays. He evidently had various schemes for essay headings and the appropriate order for these. His first page indicates some of his ideas for titles, the two columns of numbers on the left evidently referring to two different sequences:-

1	2	On Notation
2	1	Of the influence of general signs in analytical reasonings
3	3	General notions respecting Analysis (my theory of identity)
4	4	Induction
5	5	Generalisation
6	6	Analogy
8	8	Of the law of Continuity and the extreme cases of formula
9	9	Of the value of a first book
7	7	Of Artifices that of Laplace z^x

Analogous $\psi^{a,b}$

that of Landen, Euler $\dfrac{x+\overline{1-x}}{1^2} - \dfrac{x^2+\overline{1-x^2}}{2^2} + \&c.$

Analogous one

10	10	Of problems requiring new methods where the difficulty generally consists in putting it into Analytical language. Of Games. (British Museum Additional Manuscripts 37202, p. 3.)

I should point out that all the contents of 'The Philosophy of Analysis' suggests a quickly written first version. There is almost no regard for punctuation, some calculations are repeated and many of the ideas are presented roughly in note form, as the above. If we include the unnumbered 'Of games' it appears that Babbage originally proposed to write eleven essays. However, by the time he wrote the introduction, he had made amendments as follows:

The titles of the several essays which will contain the result of my enquiries are as follows   1st Of the influence of signs in mathematical reasoning   2nd Of Notation   3rd General notions respecting Analysis. These three are intended as an introduction which will explain some difficulties and introduce greater uniformity into mathematical symbols. Those which follow relate more immediately to the subject of invention and are   4th Of induction   5th Of Generalization   6 Of Analogy   7th Of the law of Continuity   8 Of the use of a register of ideas which occasionally strike the mind.   9 Of Artifices   10 I scarcely know what name I shall attach to the tenth essay as the want of an English one has hitherto compelled me to employ a very significant foreign term and to entitle it Des rapprochements. The 11 and last essay which is of some interest consists of a variety of problems requiring the invention of new modes of analysis. (*Ibid.*, p. 6.)

Comparing this list with the second one, the first six are exactly the same in title and order, apart from the first essay having its title changed slightly. In the latter five we have changes of order and the curious title 'Of the value of a first book' is replaced by 'Of the use of a register of ideas which occasionally strike the mind'. The wording of the title of number 10 (at least) is greatly changed but the intention remains just about identifiable.

However, the work in our possession, known as 'The Philosophy of Analysis', does not contain essays on all these subjects. 'On notation', as I have already mentioned, is much fore-shortened, and 'Of the influence of signs in mathematical reasoning', 'Of the law of continuity' and 'Of the use of a register of ideas which occasionally strike the mind' are missing. The first of these was the only published essay and its loss is, therefore, not serious, but it is impossible to speculate on the fate of the other two, whether they were lost or never written.

The two-page introduction supplies the purpose of these essays, and a most original one it is:

It is my intention in the following pages to attempt an examination of some of those modes by which mathematical discoveries have been made to point out some of those evanescent links which but rarely appear in the writings of the discoverer but which passing perhaps imperceptibly through his mind have acted as his unerring although his unknown guides. I would however to avoid misconception state at the outset that I have not attempted to explain the nature of the inventive faculty nor

> am I of the opinion that its absence could be supplied by rules however successfully contrived[.] all that I have proposed is by an attentive examination of the writing of those who have contributed most to the advancement of mathematical science and by a continued attention to the operations of my own mind to state in words some of those principles which appear to me to exercise a very material influence in directing the intellect in its transition from the known to the unknown. (*Ibid.*, p. 5.)

Babbage is rightly not concerned with a discussion of the psychology of invention, but rather to show by mathematical examples how discoveries have been made. He attempts in these essays to show how the process of discovery could be improved, how various methods of mathematical reasoning are performed, and how to abstract a mathematical problem from a not too obviously mathematical situation. Babbage has the very rare quality for a mathematician, of being able to think aloud, and express his thoughts with great clarity. He can explain in systematic language the elements of mathematical style and even speak to the non-mathematical.

The first two essays can be very briefly summarised. The first one 'On the influence of signs in mathematical reasoning' will be commented on in Chapter 7. Sufficient to say again that it is long and not particularly valuable. Babbage does say in the introduction, 'In presenting to the Cambridge Philosophical Society the first of a series of Essays on the *Philosophy of Analysis . . .*' (*ibid.*, p. 5) indicating that more were to follow. For reasons unknown, none of the others were published by this Society. Indeed, almost any of the others would have done their writer more credit than this one.

The second essay 'On notation' takes only a couple of pages in the British Museum manuscript. This cannot possibly have been the original essay, on the grounds of the absurd inequality in length between this and the first one, and Peacock's admitted difficulty in reading it. I think a possible explanation is that at this time Babbage was writing an article of the same title for the *Edinburgh Encyclopaedia*. This article was not published until 1830, but as the following correspondence shows, was probably written much earlier. Sir David Brewster, the editor of the *Encylopaedia*, first wrote to Babbage on 22 November 1818:

> I shall be very glad to receive from you the articles *Notation* and *Porisms*. The first of these I shall wish to be short as you

conveniently can make it, as our [readers?] are extremely irascible. It will not be wanted for a year at least, so that I shall give
you notice some time hence of the precise time when it will be
wanted, as it is impossible to calculate it at present.[5]

On 3 July 1821, Brewster wrote again. The first part of the
letter which has already been quoted, was to give reasons for not
accepting a paper of Babbage's for the *Edinburgh Journal of
Science*, and continued: 'As we are very near *Notation* you would
oblige me by sending it.'[2] Finally, on 25 February 1822, he wrote:
'I enclose a Proof of your Article on *Notation*.'[6]

As Babbage had written this scholarly and original article
(commented on in chapter 7) before early 1822, it seems most
probable that he intended this to form the major part of his essay
on the same subject in 'The Philosophy of Analysis' and this is the
explanation for its absence for the most part, in the document
that has survived.

The two pages that have survived contain new material on the
subject of notation, but their content is comparatively trivial.
Only one point is made, about inverse operations: '*When the
operation is an inverse one the sign implying it shall be the direct sign in
an inverted position*' (*ibid.*, p. 151).

It is interesting that Babbage emphasises this rule before
expanding on it. This is the same style that he used in the
Edinburgh Encyclopaedia article 'On notation', and suggests that
the two unpublished sheets were continuous with this paper as an
additional afterthought. Among the eleven published 'rules' for
notation, the one that comes closest to this is: '*Whenever we wish to
denote the inverse of any operation, we must use the same characteristic
with the index* -1'[7] His point in the *Encyclopaedia* is that '$\tan^{-1} x$' is,
for example, much neater and more suggestive than 'arc tan x'.
However, in 'The Philosophy of Analysis' (p. 151) he sees that this
is not always possible: 'It might perhaps have been productive of
some inconvenience if division as the inverse of multiplication
had been written thus x^{-1}: to have followed this principle in
forming the sign representing subtraction would have been
impossible, for writing it $+^{-1}$ is assuming its common sign in the
index.'

Commenting on the rule stated only on the unpublished sheet
he says:

In the signs representing greater and less than $>$&$<$ as well as
]&[which have also been employed for this purpose this rela-

> tion is apparent. Leibniz has on some occasion used the signs $\wedge$ and $\vee$ to signify multiplication and division which also have a similar reciprocal to each other. The sign $+$ from its symmetry does not admit of any sign for minus which can have the relation in question and had algebraic language now to take its origin there can be no doubt but that $\lfloor$ and $\rceil$ or some similarly related signs would be better constituted for the representation of plus and minus than those at present admitted. (*Ibid.*, p. 151.)

This is a change of tune from the previous rule insisting on the index -1 for inverse functions. Perhaps this new idea could be extended, for example, to trigonometric functions with tan and nat as inverses! He gives one further example which carries the logic to even greater extremes: 'Another instance is to be met with where the inverse of the operation ψx is denoted by the same sign inverted in the Journal de l'Ecole Polytechnique Cahier 15 p. 264 Laplace has made use of the notation $y = \psi x$ and $x = \phi y$' (*ibid.*, p. 151).

We next consider the third essay, 'General notions respecting analysis', which is one of the most interesting of the set. As I have said, this was a very significant period for the development of modern mathematics. Bolyai and Lobachewsky were publishing their work on non-Euclidean geometry, showing that the classical system of geometry was in no way unique, a synthetic *a priori* truth or even necessarily related to the real world, but merely one among many selected for its convenience rather than its intrinsic value. At the same time a similar revolution was taking place in algebra, and the first work which introduced the more modern concept of this subject was by Babbage's old friend and fellow-worker, George Peacock – *A Treatise on Algebra*, Cambridge, 1830. Peacock's major idea was to make a distinction between arithmetical and symbolic algebra. Previously algebra had been considered only as arithmetic, with letters and symbols replacing the numbers. Peacock rightly found this unnecessarily restrictive since arithmetic is only one branch of algebra and not the whole thing. He makes this point in his definition of algebra,

> ALGEBRA may be defined to be, *the science of general reasoning by symbolic language*.
> it has been termed *Universal Arithmetic*: but this definition is defective, inasmuch as it assigns for the general object of the science, what can only be considered as one of its applications.[8]

It will be useful for our purposes to see how Peacock comes to this conclusion in the preface to his book, and then to evaluate the degree to which he was anticipated by Babbage, roughly nine years earlier.

Peacock first indicates the way algebra had previously been considered:

> Algebra has always been considered as merely such a modification of Arithmetic as arose from the use of symbolic language, and the operations of one science have been transferred to the other without any statement of an extension of their meaning and application.[9]

He next distinguishes between the formal symbols of algebra, and their more restricted meaning when used in arithmetic:

> The imposition of the names of Addition and Subtraction upon such operations, and even their immediate derivation from a science in which their meaning and application are perfectly understood and strictly limited, can exercise no influence upon the results of a science, which regards the combinations of signs and symbols *only*, according to determinate laws, which are altogether independent of the specific values of the symbols themselves.[10]

This makes algebra a much more general subject, in which the symbols and expressions used mean no more or less than what they are defined to mean. One possible interpretation of an algebraic expression is the corresponding arithmetical one; but it is only one. It is possible, for example, to interpret signs like + and × other than in the conventional way and obtain an equally consistent system. This means 'that Arithmetic can only be considered as a Science of suggestion, to which the principles and operations of Algebra are adapted, but by which they are neither limited nor determined.'[11] Peacock then gives a more general meaning to the ' = ' sign.

> The fundamental operations of Algebra are altogether symbolical, and we might proceed to deduce symbolical results and equivalent forms by means of them without any regard to the principles of any other science; and it would merely require the introduction of some such sign as = in the place of the words *algebraical result of*, or *algebraically equivalent to*, to connect the results obtained with the symbolical representation of the operations which produce them.[12]

Having stated general principles, he then shows their implications by means of examples.

> The expression $-b+a$ is algebraically equivalent to $a-b$: if a and b bc quantities of the same kind, and if a be greater than b, then $a-b$ admits of an immediate and simple interpretation: but it is only by a reference to this second and equivalent form that we are properly enabled to interpret the first.[13]

Since algebra is a generalisation rather than a symbolisation of arithmetic, it can be used to generalise arithmetical results and this leads to the first statement of Peacock's principle:

> If m and n be whole numbers, then $ma+na = (m+n)a$, so likewise in the other, under the same circumstances $a^m \times a^n = a^{m+n}$: and inasmuch as in the one case, *the principle of the permanence of equivalent forms* would show that $ma+na = (m+n)a$, when m and n are general symbols affected with any signs whatever, so likewise in the other, the same principle, under the same circumstances, would equally show that $a^m \times a^n = a^{m+n}$; the interpretation of the meaning of particular values of the index, whether fractional or negative, is involved in this conclusion, which becomes the principle of the indices; and it becomes therefore the general principle which must not only determine the interpretation of indices when they are assumed, but must guide us conversely to the determination of the indices, which must be assumed to suit a specific interpretation.[14]

These examples make the principle abundantly clear. Results like the index law of multiplication which are demonstrably true when the indices are positive integers are used to give meaning to fractional and negative indices. If n is a positive integer, a^n means a multiplied by itself n times, but if $n = \frac{1}{2}$, then the expression $a^{1/2}$ has no meaning in this arithmetical system. But if we now apply the principle of the permanence of equivalent forms and say that the index law of multiplication is the permanent principle here, then whatever $a^{1/2}$ means, it will follow that $a^{1/2} \times a^{1/2} = a^{1/2+1/2} = a$. Consequently $a^{1/2}$ is interpreted to mean $\sqrt{a}$.

The principle is necessary to give meaning to various expressions outside the range of arithmetic:

> If n be a whole number, the existence of the equivalent series for $(1+x)^n$ is necessary, inasmuch as the operation which produces it may be completely defined; but if n be a general symbol,

we are unable to define the operation by which we pass from $(1+x)^n$ to its equivalent series which exists therefore under such circumstances, only in virtue of the principle of the permanence of equivalent forms.[15]

Mathematicians had been using this principle unconsciously for centuries. No one until the time of Babbage and Peacock had pointed out this essential distinction between arithmetic and algebra, thereby releasing the latter from its arithmetical bondage to develop in the following century as a much more general subject in its own right.

Before considering Babbage's ideas, we may summarise Peacock's thesis under the following headings:

1. Algebra had previously been thought of only as a modification of arithmetic.
2. Algebra consists of the manipulation of symbols in a way independent of any particular interpretation.
3. Arithmetic is only a special case of algebra – a 'Science of Suggestion' as Peacock put it.
4. The sign '=' is to be taken as meaning 'is algebraically equivalent to'.
5. The principle of the permanence of equivalent forms.

It can be shown that all these ideas were expressed in Babbage's essay. I have already mentioned that in his index page (British Museum Additional Manuscripts 37202, p. 3), he puts the title as 'General Notions respecting Analysis (my theory of identity)', and we will find that this theory of identity is almost exactly the same as Peacock's principle. In coming to this conclusion, Babbage traces out an argument also very similar to Peacock's.

He begins, 'Algebra appears at its first invention to have consisted of little more than the employment of a letter to represent a number to be determined by the conditions of the problem' and 'its use was therefore restricted to arithmetical enquiries' (*ibid.*, p. 41).

Here we have Peacock's first point that algebra had at first been thought of as symbolised arithmetic. However, Babbage admits that its use has been somewhat wider than this:

> To the dominion of number which algebra now possessed Descartes added that over space and large as was this addition to its empire it was perhaps scarcely less valuable as pointing out the road to other acquisitions ... The representation of time and force by means of letters and the applications of algebra to

mechanics optics and other parts of natural philosophy follows with little effort when the road was once opened. (*Ibid.*, p. 41.)

Algebra is capable of being interpreted in all these ways and yet the sum of all possible interpretations is not the complete subject. To think of algebra only in terms of such interpretations is to restrict the subject unnecessarily:

> The cause which has mainly contributed to better the language of signs may be found in this circumstance; that being in itself a method of reasoning of extreme generality it was discovered through the medium of one of its particular applications that of number; now although number itself is an abstraction and in a certain measure affects almost all the applications of analysis it is still of a far less general nature and in several instances has limited the signification of symbols by a reference to its peculiar nature. (*Ibid.*, pp. 42–3.)

The arithmetical interpretation thus imposes restrictions of an arithmetical and not an algebraical nature.

> Number is undoubtedly one of the most extensive of the subjects to which Analysis has been applied; geometry holds the second rank in point of extent neither of these can express all the relations indicated by the language of signs; the square root of a whole number for instance is not capable of numerical expression in finite terms unless it be a complete square; whilst in geometry it is always possible to express the result of such an extraction by means of a right angled triangle: if a negative sign is prefixed to the quantity whose root is to be extracted both arithmetic and geometry equally fail in executing the operation. (*Ibid.*, p. 46.)

Babbage now attempts to show how algebra can be treated as a mathematical discipline in its own right, freed from the necessity to interpret at all times:

> The object which I propose to attempt is to separate entirely Analysis or the language of signs from all its various applications rejecting from it not merely geometrical considerations but even those of number and to show that when viewed in this light it ultimately resolves itself into propositions which are purely identical or at least that the signification of every equation amounts to no more than that when all the operations which are indicated on each side are actually executed every letter which occurs on the other side will be found occurring under precisely similar circumstances on the other and in any

case any letters stand as representatives of others these latter must be substituted for them before the identity becomes apparent. (*Ibid.*, p. 44.)

This means that algebra is to be considered as the formal manipulation of symbols by prescribed rules. These rules may be interpreted but are formulated in a way independent of any interpretation, numerical or otherwise. How this can be done is shown in the following example:

> In the equation $(a+x)^3 = a^3 + 3a^2x + 3ax^2 + x^3$ the identity of the two sides is readily made manifest; nothing further is required than actually performing the two multiplications which are only expressed on the left side and it will then be found that each letter is exactly similarly situated on each and that the whole is reduced to zero by the mutual destruction of terms. Now it is important to observe that this destruction is entirely independent of the nature of the things denoted by a and x and of whatever class they may be it must equally take place. (*Ibid.*, p. 45.)

The purely operational nature of algebra is carried a stage further by an application to the theory of infinite series, and used to explain an apparently paradoxical result. From the binomial expansion $1/(1+x) = 1 - x + x^2 - x^3 + \cdots$, putting $x = 1$, the result becomes $\frac{1}{2} = 1 - 1 + 1 - 1 + \cdots$ which according to Babbage 'cannot possibly be numerically true' (*ibid.*, p. 48). However, if both sides of the equation are integrated the result $\log(1+x) = \frac{x}{1} - \frac{x^2}{2} + \frac{x^3}{3} - \cdots$ is obtained, and this time when $x = 1$ the 'true' result $\log 2 = 1 - \frac{1}{2} + \frac{1}{3} - \cdots$ is obtained. This paradox, Babbage now shows, can be resolved by reconsidering the '=' symbol. The equation $1/(1+x) = 1 - x + x^2 - x^3 + \cdots$ must be replaced by the true proposition $[1 - (-x)^{n+1}]/(1+x) = 1 - x + x^2 - \cdots (-1)^n x^n$ from which, on integration,

$$\log(1+x) + (-1)^n \int \frac{x^{n+1}\,dx}{1+x} = \frac{x}{1} - \frac{x^2}{2} + \frac{x^3}{3} \cdots (-1)^{n-1}\frac{x^n}{n}$$

which can also be written as:

$$\log(1+x) + \frac{(-1)^n x^{n+2}}{(n+2)(1+x)} + \frac{(-1)^n}{n+2} \int \frac{x^{n+2}\,dx}{(1+x)^2}$$

$$= \frac{x}{1} - \frac{x^2}{2} + \frac{x^3}{3} - \cdots + (-1)^{n-1}\frac{x^n}{n}.$$

He concludes, 'Every stage of this process is reducible to identity and is quite independent of the magnitude of x' (*ibid.*, p. 49).

This example not only illustrates again the symbolic nature of algebra, but also shows how the equality symbol should be used particularly when infinite series are concerned. In the arithmetical sense of the symbol, the equation $1/(1+x)= 1-x+x^2-x^3+\cdots$ is untrue, as the particular case $x=1$ shows; but if the symbol is taken to mean 'is derived by algebraical rules from' then the equation is true, and can also be made arithmetically true by the introduction of the remainder term $(-x)^{n+1}/1+ x$. An infinite series is then shown to exist in the algebraic and therefore the more general sense, with arithmetical interpretation under particular circumstances, e.g. when the series is convergent. This sort of analysis is a necessary preliminary to the introduction of infinite series, and obviously superior to a purely arithmetical development. As Babbage puts it:

> Such an explanation by no means banishes the use of infinite series, it merely delays their introduction a few stages later and the slight addition which is thus made to the length of the operations is more than counterbalanced by the logical precision which it introduces and by removing a source of error to which our reasonings would otherwise be open. (*Ibid.*, p. 50.)

We have now found arguments equivalent to the first four of Peacock's. The fifth, tantamount to the principle of the permanence of equivalent forms, is to be found on p. 43:

> $x^a \times x^b = x^{a+b}$. This equation which is true for whole numbers does not necessarily subsist when the exponents are fractions nor does the latter follow from the original assumptions; still less does it subsist necessarily when a and b are imaginary quantities.
>
> In order to give meaning to such expressions we must have recourse to a new definition and to avoid ambiguity it is extremely convenient that the new definition should include the old one as a particular case this has been accomplished by assuming the equation $x^a \times x^b = x^{a+b}$ as the definition of the operations denoted by the application of the exponents to the quantity x.

Apart from giving this principle a different name, Babbage's idea seems almost identical to Peacock's.

We have now established a quite remarkable similarity in the thinking of two contemporary mathematicians, and we have the evidence of Peacock's letter that he had read Babbage's work nine years prior to his own publication. However, there is not a shred of additional evidence to support a charge of plagiarism or even collusion. Peacock's subsequent letters to Babbage in the British Museum collection are all short, trivial and friendly. The first one after the publication of *A Treatise on Algebra* is dated 18 June 1833, has no mathematical content at all, and ends:

> I will get your rooms ready . . . if you come by the Telegraph. Believe me My dear Babbage Most truly yours
> G. Peacock.[16]

The remaining nine letters spread over the next seventeen years are about subscription lists, academic politics, examination results, booking rooms and proposing votes of thanks. The only one which indicates anything at all about their personal relationship was written by Peacock in 1850: 'I was very glad to see your handwriting again as it is so long since I have heard from you.'[17]

In view of such lack of evidence it seems idle to put forward any theory to account for the astonishing similarity of Babbage's unpublished and Peacock's published work. It might even be possible that Peacock took in Babbage's ideas unconsciously, and with full integrity turned them out as his own a few years later. In any case, Babbage was too busy working on his computer and reforming British science generally to worry very much about the priority of any of his inventions in pure mathematics. Sufficient to say that in the discovery of so called 'modern algebra' Babbage's work in 1821 was substantial.

He also has some interesting remarks to make about the foundations of differential calculus in this chapter. He is quite sure that the vexed problem of finding a logical foundation for such a branch of mathematics has been solved once and for all, for on p. 42 he says: 'The masterly view of that subject which was taken by Lagrange whilst it rested the calculus on a foundation not to be contested has at the same time furnished an example which will have no inconsiderable weight in guiding other reforms.'

Babbage is obviously referring to J. L. Lagrange's *Théorie des Fonctions Analytiques*, Paris, 1797, in which the writer attempted to establish formally the successive derivatives of a function as the coefficients in the corresponding Taylor series expansion, a proof of this series for all functions being given algebraically. It is now known that such a general result cannot be proved by algebra alone and considerable doubt was cast on Lagrange's method even by his contemporaries. Babbage, writing of his early life at Cambridge in *Passages from the Life of a Philosopher*, London, 1864, admits that he learnt the Leibnizian notation of the d's for calculus from Robert Woodhouse's *Principles of Analytical Calculation*, Cambridge, 1803. If he had studied the important preface to this work he would have found a convincing refutation of Lagrange's theory, showing circular arguments, the shifting of hypotheses and the blatant use of limits which Lagrange was specifically trying to avoid.

Babbage describes elsewhere in this passage 'The Philosophy of Analysis' the essential difficulty of the process of calculus, in that no matter what method is used, be it fluxions, differentials, infinitesimals, indivisibles or ultimate ratios, somewhere or other consideration must move from the finite to the infinite. This not only presents a problem but gives greater pleasure on solution for

> An evident satisfaction in verifying by experiment the truth of those conclusions [at] which we have arrived after having traversed the obscure regions of infinity.
>
> As no sagacity can enable us to elude that step which the very constitution of the subjects of our enquiry renders necessary it becomes of some importance to determine rightly the most proper place for its introduction. (*Ibid.*, p. 51.)

The problem then faced by formulators of the calculus is to find the right place to start from, so that this inherent difficulty of passing from the finite to the infinite is overcome. Babbage has his own suggestion when he says on p. 54: 'That which I have made choice of is $d(xy) = x \, dy + y \, dx$ which combined with $d(x + y) = dx + dy$ will produce all the others.'

Taking this formal definition of the operator d, it is quite easy to establish results as Babbage did, such as $d(x^n) = nx^{n-1} \, dx$, for n not only a positive integer, but also a negative integer or a fraction. One can establish the formal rules for calculus such as the quotient rule and differentiation of a function of a function. The passage from the finite to the infinite appears to arise here when attempting to establish results like Taylor's series so that functions other than simple polynomials can be dealt with. Babbage's two rules taken in conjunction with his conception of infinite series referred to in this chapter can take us a long way formally into the calculus, but it seems impossible to derive the complete theory without introducing somewhere a process equivalent to the use of limits.

The next three chapters on induction, generalisation and analogy form a group together. The lines of demarcation between these three is not always apparent but is indicated in the chapters by examples rather than definitions. Babbage is concerned here with principles of mathematical discovery rather than justification, and his ideas and examples are useful for giving an indication of how a mathematical mind often works.

According to F. Cajori's *A History of Mathematics*, New York, 1919, Augustus De Morgan was the first to introduce the name 'mathematical induction' in an article for the *Penny Cyclopaedia* in 1838.[18] The term 'induction' had been used by John Wallis in *Arithmetica Infinitorum*, London, 1656, in the same sense as the word is used in natural science. To quote from Cajori's book: 'From Wallis to De Morgan, the term 'induction' was used occasionally in mathematics, and in a double sense, (1) to indicate incomplete inductions of the kind known in natural science, (2) for the proof from n to $n + 1$. De Morgan's 'mathematical induction' assigns a distinct name for the latter process.[18]

Babbage uses this word in both senses in his chapter 'Of induction' for 'The Philosophy of Analysis'. He demonstrates the usefulness of (2) and the total unreliability of (1) while admitting that sometimes it is useful for suggesting results. Dealing first with (1), he says (*ibid.*, p. 56):

> In mathematical enquiries the method of induction is said to be made use of when by examining a few particular cases of a theorem we conclude the Truth of some general law such an instance occurred in investigating the binomial theorem for whole numbers; a few of the first powers give
> 1st $1 + x$
> 2nd $1 + 2x + x^2$
> 3rd $1 + 3x + 3x^2 + x^3$
> 4th $1 + 4x + 6x^2 + 4x^3 + x^4$
> 5th $1 + 5x + 10x^2 + 10x^3 + 5x^4 + x^5$.

From these instances the general rule for the expansion of $(1 + x)^n$ where n is a positive integer, is deduced. Some sort of induction, he asserts, is needed to change from 'an highly probable to a necessary truth' (*ibid.*, p. 56). The nature of the 'necessary truth' thus obtained is an example of the great economy of mathematical generality in that 'The foundation of *one single instance* determines the necessary truth of a series of operations extending indefinitely' (*ibid.*, p. 56). However, the 'one single instance' is much more certainly established by the use of induction method (2) rather than (1) and he follows this remark by showing how the normal mathematical induction method works to prove the expansion for $(1+x)^n$.

From the expansion of $(1+x)^n$ for positive integral values of n, Newton guessed rightly the infinite series for $(1+x)^n$ when n is any rational number. Babbage points out that this is not to be taken as an induction, but a generalisation, making the first of his distinctions between these ideas.

Turning to number theory, he says, 'The method of induction has been employed in the theory of numbers perhaps more frequently than in any other branch of analysis and by a singular coincidence there is none in which it so frequently leads to error.' As evidence for this he quotes Fermat's mistake in asserting that any number of the form $2^{2^n}+1$ was prime. 'The number of instances on which his induction was grounded could only be four since Euler has shown that it fails in the fifth number' (*ibid.*, p. 57). In fact $2^{2^5}+1 = 4\ 294\ 967\ 297 = 6\ 700\ 417 \times 641$.

Babbage's remark about induction being required to proceed from probable to necessary truths associates the concepts of induction and probability together. He now gives an analysis of the idea of probability.

> When the term probable is applied to any truth which is established by such evidence as that we are now considering, it should be observed that it does not apply to the nature of the proposition which is in all cases *necessarily* true or *necessarily* false but it applies to the arguments which induce us to believe one of these to be the case rather than the other; according to the number and quality of these arguments we decide and it is obvious that they may vary through all degrees from bare probability up to a conviction beyond even moral certainty. (*Ibid.*, p. 59.)

It is certainly correct to apply the term 'probable' not to the conclusion which is being tested but to the nature of the evidence being presented. However, it is extremely unlikely that in analytical reasoning it can ever be possible to assign a probability value to the 'number and quality' of the supporting arguments. However many particular cases are found in support of a hypothesis, their number and quality do not prove the result and do not even make it probable. From a finite number of cases, infinitely many differing hypotheses are suggested, and Babbage proceeds to give two interesting examples which show the inadequacy of this method of induction, and seem to refute his assertion about probability.

He first considers the expression '$x^2 + x + 41$' (*ibid.*, p. 63). It can be demonstrated that giving x any integral value between 1 and 40, $x^2 + x + 41$ will be a prime number, and proceeding by induction one might think that the result was generally true for any positive integer. But the expression is obviously not prime for $x = 41$, and incidentally, to correct Babbage, for $x = 40$. It is clear from this example that the fact of the first thirty-nine numbers all agreeing with this hypothesis does not make it true, or even probable. In fact, from the evidence the probability of the general result being true, since it depends on all possible cases, is precisely zero.

The second example is an even more convincing demonstration, because whereas the first one breaks down at a comparatively low number ($x = 40$), this one does not break down until the number n is reached on the positive integral scale, and n may be as large as we like.

> I shall show that even in a very simple case two expressions may coincide with each other for any given number of terms and yet fail at some other.
>
> Let ^nSi denote the sum of the ith powers of the nth roots of unity divided by n then $i + {}^nSi$ will be the general term of the series $1, 2, 3, \ldots n-1, n+1, n+1, n+2, n+3, \ldots$ and this coincides with the supposition that the general term is i for the first $n-1$ cases but it fails in the nth. (*Ibid.*, p. 63.)

To explain this, the nth roots of unity will be the solutions of the equation $x^n - 1 = 0$. It can be shown that these roots are in geometric progression and can be expressed as $1, w, w^2, \ldots w^{n-1}$; $w = \cos(2\pi/n) + i \sin(2\pi/n)$. Also, by the theory of equations, $1 + w + w^2 + \cdots + w^{n-1} = 0$. If we consider the sum of the ith powers of these n roots, we obtain $1^i + w^i + w^{2i} + \cdots + w^{(n-1)i}$. No two of these are equal unless n is a factor of i, since if two of them, say w^{ij} and w^{ik} are equal, then $w^{i(j-k)} = 1$ and either $j - k = 0$ or is a multiple of n, or i, which is non-zero, is a multiple of n. Since $j \neq k$ and both j, k lie between 0 and $n-1$, the only circumstance under which any two of these could be equal, is for i to be a multiple of n.

If i is not a multiple of n, the $1^i, w^i, w^{2i}, \ldots w^{(n-1)i}$ are all distinct, and a re-arrangement of $1, w, w^2, \ldots w^{n-1}$. It follows that $1^i + w^i + \cdots + w^{(n-1)i} = 1 + w + \cdots + w^{n-1} = 0$. Then $i + {}^nSi = i$.

Otherwise, if i is a multiple of n,

$$1^i = w^i = \cdots = w^{(n-1)i} = 1 \text{ and } 1^i + w^i + \cdots + w^{(n-1)i} = n.$$

In this case, $i + {}^nSi = i + 1$. We have, therefore, a most ingenious example of a general rule which follows the series $1, 2, 3, \ldots$ up to $n - 1$ for n as large as we please, but which differs from the series at n. This is an excellent counter-example to show that induction type (1) can never be trusted in mathematics, for no matter how many times a rule may seem to be confirmed, the rule may still be false.

While this method is useless as a proof, it can still be helpful in suggesting results. Babbage supplies one example from his own work although he did not often use this type of reasoning. 'Sensible that in my own case whenever I have been successful in carrying my enquiries a few steps out of the beaten path I have been more indebted to generalisation and to analogical reasoning than to the processes of induction. I shall not venture to offer many observations on which my experience is very limited' (*ibid.*, p. 49).

The example is that in continually multiplying the number 7 by itself, he observed that the resulting figure ended with the digits 01 in the 4th, 8th, 12th, 16th, ... places, and 001 in the 20th, 40th, ... places. This suggested a general rule which he proceeded to prove analytically.

Taking any positive integer a, let i be the value of x for which the expression $(7^x - 1)/10^a$ is a whole number. Then $(7^i - 1)/10^a = w$ or $7^i = 10^a w + 1$. Raising each side to the power v, $7^{iv} = (10^a w + 1)^v$: and putting $iv = x$, $(7^x - 1)/10^a = [(10^a w + 1)^v - 1]/10^a$. Expanding $(10^a w + 1)^v$ by the binomial series, $(10^a w + 1)^v - 1$ is clearly divisible by 10^a so that for all positive integral values of v, $(7^x - 1)/10^a$ will be a whole number for $x = iv$.

The chapter ends with an example from Wallis's work.

> Few works afford so many examples of pure and unmixed induction as the Arithmetica Infinitorum of Wallis and although more rigid methods of demonstration have been substituted by modern writers this most original production will never cease to be examined with attention by those who interest themselves in the history of analytical science or in examining

those trains of thought which have contributed to its perfection. (*Ibid.*, p. 66.)

The example given is:

$$\frac{0+1}{1+1} = \frac{1}{2} = \frac{1}{3} + \frac{1}{6}$$

$$\frac{0+1+4}{4+4+4} = \frac{5}{12} = \frac{1}{3} + \frac{1}{12}$$

$$\frac{0+1+4+9}{9+9+9+9} = \frac{14}{36} = \frac{7}{8} = \frac{1}{3} + \frac{1}{18}$$

$$\frac{0+\ 1+\ 4+\ 9+16}{16+16+16+16+16} = \frac{30}{80} = \frac{3}{8} = \frac{1}{3} + \frac{1}{24}$$

$$\frac{0+\ 1+\ 4+\ 9+16+25}{25+25+25+25+25+25} = \frac{55}{150} = \frac{11}{30} = \frac{1}{3} + \frac{1}{30}$$

'If therefore the number of terms be augmented indefinitely this ratio becomes correct within any assignable limits' (*ibid.*, p. 66).

This particular sequence of results is not surprising since it can be proved without much difficulty even with the resources available to Wallis and his contemporaries that the general expression

$$\frac{0^2 + 1^2 + \cdots + n^2}{n^2 + n^2 + \cdots + n^2} = \frac{\frac{1}{6}n(n+1)(2n+1)}{n^2(n+1)} = \frac{1}{3} + \frac{1}{6n}.$$

Wallis's work is of great interest, however, as an early example of a potential procedure to a limit by purely arithmetical means.

It is easy to deplore the logic of this elementary type of induction but hard to deny its value in the development of mathematics. The method can easily be shown to be unsound, but many correct results have been found through it, to be proved later by superior techniques.

The chapter 'Of generalization' (*ibid.*, pp. 68–82) is not nearly so interesting and it is difficult to see why it should have been included at all. Babbage admits that it is hard to draw the line between induction (in the first sense) and generalization. He says:

> In fact the term induction has been applied to two processes very distinct in themselves either of which may be and frequently is employed by itself.
> The first part of the process consists in comparing together a number of particular instances and determining some property

or character which is common to them all[.] this is exactly
analogous to that process by which we arrive at general terms in
language and[,] to the mental operation by which it is accomplished[,] the name of abstraction has been given.

In as far then as generalization and abstraction are synonym-
ous the first part of the inductive process can only be carried on
by their aid. (*Ibid.*, pp. 71–2.)

The second process he takes as equivalent to justification.
Generalization, then, is almost exactly the same as the process of
induction described in the previous chapter. There is a difference, as the example given of Newton's infinite binomial series
generalised from cases of finite series indicates, but one which is
difficult to define. The indirect method of definition is continued
on p. 74:

The line which separates a process of generalization from one
of analogy is rather more clearly marked than that which
distinguishes it from induction; analogy transfers principles
and modes of operation to different classes whilst generalization embraces in its grasp all those species which are comprised
in some one class[.] neither of them afford proof of the truth of
the conclusions to which they lead but when these have been
demonstrated by other means it will be found that those which
have been derived from a principle of generalization always
contain in them the theorems from which they derived their
origin; whilst those who have been suggested to us by analogy
do not always include but only resemble their prototypes.

This is a perfectly clear distinction. When a result has been
generalised, the new result includes the former as a special case,
whereas in the procedure by analogy there is not necessarily any
such relation. A further distinction is given on p. 75: 'Generalization is frequently the result of an enquiry induction and analogy
are the means by which it is accomplished.'

The examples given to illustrate the process of generalization
are very scanty and it will suffice to give only one. Babbage quotes
(*ibid.*, p. 76) the example given by Laplace that the sum of the
squares of the coefficients of the binomial series $(1 + x)^n$, namely
$1 + (n/1)^2 + (n \cdot n - 1/1 \cdot 2)^2 + \cdots$ is equal to the coefficient of the
middle term of the expansion $(1 + x)^{2n}$. This result is extended to
show that if $(1 + ax + x)^{2n} = A + A_1 x + A_2 x^2 + \cdots$, then $A^2 + A_1^2 +$
$A_2^2 + \cdots =$ the coefficient of the middle term of the expansion
$(1 + ax + x^2)^{2n}$. This can now be generalised to say that

if $(1 + ax + bx^2 + \cdots + bx^{k-2} + ax^{k-1} + x^k)^n = A + A_1x + A_2X^2 + \cdots$,
then $A^2 + A_1^2 + A_2^2 + \cdots =$ the coefficient of the middle term of the
expansion $(1 + ax + bx^2 + \cdots + bx^{k-2} + ax^{k-1} + x^k)^{2n}$.

The next chapter 'Of Analogy' (*ibid.*, pp. 83–91) is also incomplete in that the ideas and examples given in the text actually have less content than those given in an earlier paper, 'Observations on the analogy which subsists between the calculus of functions and other branches of analysis'.[19] The chapter has only two major examples, one which is largely the same as a similar one given in the above paper and the other which was also included in *Examples of the Solutions of Functional Equations*, Cambridge, 1820.

The first of these ('The Philosophy of Analysis', pp. 84–9) compares the calculus of functions by analogy to differential equations and difference equations. The differential equation $dy + Py\,dx = Q\,dx$ can be solved by multiplying both sides by the integrating factor $e^{-\int P dx}$, and Babbage wondered if difference equations and similar problems from the calculus of functions could be solved by a similar device. To solve the difference equation (*ibid.*, p. 84), $\Delta y + Py = Q$, put $y = uz$ so that $\Delta y = u\Delta z + z\Delta u + \Delta u\Delta z$ and the equation becomes $u\Delta z + z\Delta u + \Delta u\Delta z + Puz = Q$. Now choose u in such a way that $z\Delta u + Puz = 0$ or $\Delta u + Pu = 0$. The equation $\Delta u + Pu = 0$ can be integrated immediately to a particular solution $u = (1 - P)^x$. This leaves $u\Delta z + \Delta u\Delta z = Q$ or $\Delta z = Q/(u + \Delta u)$. On summation, this means that $z = \sum[Q/(u + \Delta u)]$ and, since $u = (1 - P)^x$, $u + \Delta u = (1 - P)^{x+1}$. The general solution is, therefore, $y = (1 - P)^x \sum[Q/(1 - P)^{x+1}]$. He concludes (*ibid.*, p. 85) 'Analogy which doubly suggested this plan induced me to try the effect of some similar artifice for the solution of functional eq^{ns}.'

The particular example he tackles (*ibid.*, p. 88) is the equation

$$\psi x + x^4 \psi(1/x) = x^2 \frac{1 + 2ax + x^2}{1 - 2ax + x^2}. \tag{1}$$

The technique is to observe that the symmetrical equation $fx \cdot \psi x + f\alpha x \cdot \psi\alpha x = f_1 x$ where $\alpha^2 x = x$, may be reduced to $\psi_1 x + \psi_1 \alpha x = f_1 x$ by the substitution $\psi_1 x = fx \cdot \psi x$ and this can be solved as $\psi_1 x = fx\phi x/(\phi_1 x + \phi_1 \alpha x)$. Therefore, to make (1) symmetrical, multiply both sides by the function ϕx, so that the equation becomes

$$\phi x \cdot \psi x + x^4 \phi x \psi(1/x) = x^2 \frac{1 + 2ax + x^2}{1 - 2ax + x^2}\phi x.$$

It is required that $x^4\phi x$ shall be the same function of $1/x$ as ϕx is of x. This means that ϕx must satisfy the subsidiary functional equation $\phi(1/x) = x^4\phi x$. This equation is known to have a general solution

$$\phi x = x^{-2}X\left(\bar{x}, \frac{\bar{1}}{x}\right),$$

and since we only require a particular simple solution for ϕx, it is

sufficient to take $X\left(\bar{x}, \frac{\bar{1}}{x}\right) = 1$ and $\phi x = (1/x^2)$.

Equation (1) now becomes

$$(1/x^2)\psi x + x^2\psi(1/x) = \frac{1 + 2ax + x^2}{1 - 2ax + x^2}.$$

Putting $\psi_1 x = (1/x^2)\psi x$, this becomes

$$\psi_1 x + \psi_1(1/x) = \frac{1 + 2ax + x^2}{1 - 2ax + x^2}.$$

The complete solution is then

$$\psi_1 x = (1/x^2)\psi x = \frac{1 + 2ax + x^2}{1 - 2ax + x^2} \cdot \frac{\phi x}{\phi x + \phi(1/x)},$$

where ϕx is an arbitrary function.

The analogy here is by suggestion rather than imitation. The suggestion from the theory of differential or difference equations is that by multiplying through by a term which becomes the 'integrating factor', the left-hand side may be integrated. Applying this idea to the calculus of functions leads not to an equivalent 'integration' but to a standard equation of simpler kind. Babbage admits this limitation when he adds,

> Should other modes of calculation be here discovered whenever it becomes necessary to perfect the inverse branches we may look with some confidence to the assistance of this principle: the difficulty of its application will consist in discovering something in the new calculus which has an analogy to the complete differential. (*Ibid.*, pp. 89–90.)

The second and last example in this chapter of Babbage's is again on functional equations, inspired apparently by Laplace's solution of the difference equation $1 + b(u_{x+1} - u_x) + u_x u_{x+1} = 0$ by the separation of u_x and u_{x+1}. Babbage observes that if by differentiation, we eliminate the constant b from the equation, $1 + b(u + v) - uv = 0$, there remains $du/(1 + u^2) + dv/(1 + v^2) = 0$. Now let $b = fx$ where fx is any function of x symmetrical with respect to x and αx, and let $u = \psi x$, $v = \psi \alpha x$. The equation $1 + fx(\psi x + \psi \alpha x) - \psi x \psi \alpha x = 0$ becomes, on the elimination of fx,

$$\int \frac{d\psi\, x}{1 + (\psi x)^2} + \int \frac{d\psi\, \alpha x}{1 + (\psi \alpha x)^2} = a.$$

It can actually be shown that

$$a = \int \frac{df\, x}{1 + (fx)^2} = \tan^{-1} fx.$$

Putting

$$\psi_1 x = \int \frac{d\psi\, x}{1 + (\psi x)^2},$$

the equation takes the form $\psi_1 x + \psi_1 \alpha x = a$ and this has general solution $\psi_1 x = a\phi x/(\phi x + \phi \alpha x)$, where ϕx is arbitrary. Since $\psi_1 x = \tan^{-1} \psi x$, on integration the general solution of the original equation is $\psi x = \tan[a\phi x/(\phi x + \phi \alpha x)]$.

We cannot profess to have learnt very much about Babbage's inventiveness from the chapter 'Of Analogy', apart from illustrations of his usual facility in extracting hints of methods to solve functional equations from his great knowledge of techniques used in other branches of mathematics.

The next chapter, 'Of artifices' (*ibid.*, pp. 92–123), is concerned with a display of such techniques. He begins, 'Under the general term *artifices* may be included all those contrivances by which a difficulty is either eluded, or overcome by indirect means. Almost all the advances in pure analysis which have gradually extended its boundaries have been more or less indebted to this as their source.' The chapter does not contain much original material by Babbage, but gives us a fascinating sequence of mathematical stratagems used to solve a variety of problems. Apart from their own interest, they provide useful material concerning the sources of Babbage's influences. He next enlarges on his point that mathematical progress depends more on following unexplored ideas than in pursuing a generally accepted method of reasoning.

> Detached and isolated as are those contrivances by which the
> limits of the science become extended it might at first appear

that they were destined for ever to remain monuments of the skill and ingenuity of their inventor; but the mind which has achieved conquests such as these rarely rests satisfied with the solution of the question which first excited its energies, endowed with strong habits of generalization kindred and collateral questions cannot fail to continually present themselves and the success which has crowned the first attempt stimulates to further exertions at the same time that it renders their result more certain: the original artifice modified and adapted to the new difficulties which occur no longer stands isolated and alone[;] connected with a series of operations tending to one point it forms part of system and an attentive examination of its structure and mode of operation may enable us to discover some more general expression which comprehends in its wide grasp but at opposite extremes artifices with whose use we are familiar as well as these the newly acquired fruits of our perseverance. (*Ibid.*, pp. 92–3.)

On attempting to clear through this passage written without much thought for punctuation and grammar, the major point seems to be that a mathematician having discovered a new artifice for tackling a particular problem, can never leave this technique as it stands, but strives to generalise and to develop it in such a way that previously accepted methods are seen as particular cases of this new one. This point is illustrated by examples in the text. The first one starts with a method for solving the integral,

$$\int \frac{\mathrm{d}x}{x} \log (1-x).$$

Expanding $\log (1-x)$ as an infinite series and integrating term by term, it is found that

$$\int \frac{\mathrm{d}x}{x} \log (1-x) = \frac{x}{1^2} + \frac{x^2}{2^2} + \frac{x^3}{3^2} + \cdots$$

Actually, the arbitrary constant is omitted, and the right-hand side should be the negative of what is displayed here. As these two errors do not affect the major principle of the argument, we need pursue them no further.

Substituting $1-x$ for x, the result

$$\int \frac{-\mathrm{d}x}{1-x} \log x = \frac{1-x}{1^2} + \frac{\overline{1-x}^2}{2^2} + \frac{\overline{1-x}^3}{3^2} + \cdots$$

is obtained. Now adding the two integrals on the left gives us an exact differential to integrate. In fact,

$$\int \frac{dx}{x} \log(1-x) + \int \frac{-dx}{1-x} \log x = \log x \log(1-x) + C.$$

$$\log x \log(1-x) + C = \frac{x + \overline{1-x}}{1^2} + \frac{x^2 + \overline{1-x}^2}{2^2} + \cdots,$$

putting $x = 0$, and assuming presumably that, $x \overset{\text{Lt}}{\to} 0$, $\log x \log(1-x) = 0$, we obtain $C = 1/1^2 + 1/2^2 + 1/3^2 + \cdots = \pi^2/6$. Next, putting $x = \frac{1}{2}$, so that $x = 1 - x$,

$$(\log \tfrac{1}{2})^2 + \pi^2/6 = 2(1/1^2 \cdot 1/2 + 1/2^2 \cdot 1/2^2 + 1/3^2 \cdot 1/2^3 + \cdots)$$

Now,

$$\int_0^{1/2} \frac{dx}{x} \log(1-x) = 1/1^2 \cdot 1/2 + 1/2^2 \cdot 1/2^2 + 1/3^2 \cdot 1/2^3 + \cdots$$

and

$$\int_0^{1/2} \frac{dx}{x} \log(1-x) = (\log 2)^2/2 + \pi^2/12.$$

This result is absurd since the left is negative and the right positive. Correcting the mistake made at the beginning, the solution should be

$$\int_0^{1/2} \frac{dx}{x} \log(1-x) = (\log 2)^2/2 - \pi^2/12.$$

Babbage now sees (*ibid.*, p. 98) that this method can be generalised to solve a variety of problems in integration. 'The success of the artifice depends on this circumstance that in an integral

$$\int \frac{d\phi\, x}{dx} dx \times \psi x$$

which we are unable to integrate such a function of x is substituted that it is changed into

$$\int \frac{d\psi\, x}{dx} dx \times \phi x$$

the sum of these two being a complete differential.'

Starting from the integral

$$\int \frac{\mathrm{d}\psi\, x}{\mathrm{d}x}\, \mathrm{d}x \times \psi x$$

and substituting αx for x gives

$$\int \frac{\mathrm{d}\phi\, \alpha x}{\mathrm{d}\alpha\, x}\, \mathrm{d}\alpha\, x \times \psi \alpha x.$$

It is now required that the sum of these two should be of the form $\int (v\, \mathrm{d}u + u\, \mathrm{d}v)$. This means that

$$\frac{\mathrm{d}\phi\, x}{\mathrm{d}x}\, \mathrm{d}x = \frac{\mathrm{d}\psi\, \alpha x}{\mathrm{d}\alpha\, x}\, \mathrm{d}\alpha x$$

and

$$\frac{\mathrm{d}\phi\, \alpha x}{\mathrm{d}\alpha\, x}\, \mathrm{d}\alpha x = \frac{\mathrm{d}\psi\, x}{\mathrm{d}x}\, \mathrm{d}x.$$

Integrating both of these, $\phi x = \psi \alpha x$ and $\phi \alpha x = \psi x$, so that $\psi x = \psi \alpha^2 x$.

This means that an integral of the form

$$\int \frac{\mathrm{d}\psi\, \alpha x}{\mathrm{d}\alpha\, x}\, \mathrm{d}\alpha\, x \times \psi x$$

can be evaluated by the substitution of αx for x, provided that $\alpha^2 x = x$.

In particular, if

$$\int \frac{\mathrm{d}\psi\, \alpha x}{\mathrm{d}\alpha\, x}\, \mathrm{d}\alpha\, x \times \psi x = Ax + Bx^2 + Cx^3 + \cdots$$

where $\alpha^2 x = x$, then

$$\psi x \times \psi \alpha x + C = A(x + \alpha x) + B[x^2 + (\alpha x)^2] + \cdots$$

He next considers a problem introduced by Fagnani on the computation of the lengths of arc of ellipses and other curves. This gave rise to an equation of the form

$$\int \frac{\mathrm{d}x}{\sqrt{(1 - x^4)}} = \int \frac{\mathrm{d}y}{\sqrt{(1 - y^4)}}.$$

One solution is obviously $y = x$. A more complicated solution is $y = -\sqrt{[(1 - x^2)/(1 + x^2)]}$. Babbage does not state how this was found. We can only assert that it does check. This equation is equivalent to $x^2y^2 + x^2 + y^2 - 1 = 0$, which is symmetrical in x and y as would be suggested by the nature of the equation to be solved. This particular solution suggests that a trial be made of the expression $x^2y^2 + a(x^2+y^2) + bxy + c = 0$, and from this the complete integral $x^2 + y^2 + c^2x^2y^2 = c^2 + 2xy\sqrt{(1-c^4)}$ is deduced.

Turning to the subject of recurring functions again, (*ibid.*, p. 103), Babbage considers the formulae concerning circular functions, $\sin\theta = \sqrt{[1-(\cos\theta)^2]}$, $\cot\theta = 1/\tan\theta$ and $\tan(\theta-\alpha) = (\tan\theta - \tan\alpha)/(1-\tan\theta\tan\alpha)$. He makes the point that if these three are written in the form $y = \sqrt{(1-x^2)}$, $y = 1/x$ and $y = (a-x)/(1-ax)$, then all are cases of $y = ax$, where $\alpha^2x = x$.

A moment's reflection leads to the thought that these are all obvious consequences of definitions, but the observation is not without its tautological interest.

After a consideration of various Diophantine problems in which particular and not general solutions are found, so that the content, from our point of view, is limited, Babbage turns to a problem on partial differential equations the solution of which depends on a concealed zero being located. The problem, solved by Poisson, (*ibid.*, pp. 107–8), concerns the equation $P = (rt - s^2)^a Q$, where p, q, r, s, t are taken for the operators

$$\frac{\partial}{\partial x}, \frac{\partial}{\partial y}, \frac{\partial^2}{\partial x^2}, \frac{\partial^2}{\partial x\,\partial y}, \frac{\partial^2}{\partial y^2},$$

respectively, Babbage actually using straight d's in his notation. Q is any function of x, y; a is any positive quantity; and P any function of p, q, r, s, t, homogeneous with respect to the last three and operating presumably on Q. The solution is obtained by supposing that $q = f(p)$. Then differentiating partially with respect to x, $s = rf^1(p)$ and with respect to y, $t = sf^1(p)$. Eliminating $f^1(p)$, $rt - s^2 = 0$. This means that the right-hand side of the original equation is zero, and so, therefore, is the left. The remaining equation $P = 0$ can now be solved by dividing through by an appropriate power of r, since it is homogeneous in r, s and t. For since

$$\frac{s}{r} = f^1(p) \quad \text{and} \quad \frac{t}{r} = [f^1(p)]^2$$

there remains only a differential equation relating p, $f(p)$ and $f^1(p)$. This can be solved to determine the form of the function f, and then $q = f(p)$ becomes a first order partial differential equation, which can usually be solved.

It seems that Babbage has not formulated the problem strictly enough, and that, as stated, the division by a power of r will not normally eliminate r from the equation. However, a more precise definition of the question will overcome this difficulty and the major principle remains valid.

Two artifices follow, both concerned with finite differences, and attributed to Laplace. The first of these (*ibid.*, pp. 109–11) attempts to find a formula for the nth difference of a function, $\Delta^n u$ in terms of the successive derivatives of the function. Suppose it is assumed that

$$\Delta^n u = \frac{d^n u}{dx^n} h^n + A \frac{d^{n+1} u}{dx^{n+1}} h^{n+1} + \cdots,$$

to determine the coefficients $A', A'', \ldots$, it being taken that $u = f(x)$, $\Delta u = f(x+h) - f(x)$, $\Delta^2 u = \Delta u(x+h) - \Delta u(x)$, etc. Since the result is general, it will be true for any particular function, u. Suppose then that $u = e^x$ (the exponential function). In this case

$$\frac{d^m u}{dx^m} = e^x$$

for all m. Also

$$\Delta u = e^{x+h} - e^x = e^x(e^h - 1),$$

$$\Delta^2 u = e^{x+h}(e^h - 1) - e^x(e^h - 1) = e^x(e^h - 1)^2$$

and in general

$$\Delta^m u = e^x(e^h - 1)^m.$$

The formula then becomes:

$$e^x(e^h - 1)^n = e^x h^n + A' e^x h^{n+1} + \cdots,$$

or

$$(e^h - 1)^n = h^n + A' h^{n+1} + \cdots$$

The coefficients $A', A'', \ldots$ then become those of the expansion of $(e^h - 1)^n$. Putting $n = -1$:

$$(e^h - 1)^{-1} = h^{-1} + A' + A'' h \cdots + A^{(k)} h^{n-1} + \cdots$$

Multiplying by h

$$\frac{h}{e^h-1} = 1 + A'h + \cdots + A^{(k)}h^k + \cdots$$

Differentiating k times with respect to h,

$$\frac{d^k}{dh^k}\left(\frac{h}{e^h-1}\right) = k!A^{(k)} + \frac{(k+1)!}{1!}A^{(k+1)}h + \cdots$$

This immediately gives a general formula for $A^{(k)}$ as:

$$A^{(k)} = \frac{1}{k!}\frac{d^k}{dh^k}\left(\frac{h}{e^h-1}\right)$$

where $h = 0$ after the operation.

The other artifice by Laplace concerns the solution of the difference equation $1 - B(x_{n+1}-x_n) + x_n x_{n+1} = 0$ by the elimination of B, to which reference had been made on an earlier page. The general theory, explained by Babbage (*ibid.*, p. 113), is that if on differentiation of any difference equation, something of the form $\phi(x_n)\,dx_n = \psi(x_{n+1})\,dx_{n+1}$ is obtained, then it can be integrated as $\int \psi(x_{n+1})\,dx_{n+1} - \int \phi(x_n)\,dx_n = a$. If, moreover, ψ and ϕ are the same function, then the equation reduces simply to $\Delta \int \phi(x_n)\,dx_n = a$.

This is illustrated by the difference equation $1 - B(x_{n+1}-x_n) + x_n x_{n+1} = 0$. Differentiating and eliminating B gives

$$\int \frac{dx_{n+1}}{1+x_{n+1}^2} - \int \frac{dx_n}{1+x_n^2} = a.$$

This means that

$$\Delta\left[\int \frac{dx_n}{1+x_n^2}\right] = a,$$

so that

$$\int \frac{dx_n}{1+x_n^2} = an + b,$$

where a and b are constants. Integrating, $\tan^{-1} x_n = an + b$, so that the solution $x_n = \tan(an+b)$ is obtained.

Babbage points out that ϕ and ψ need not be identical for a solution to be found for

$$\int dx_{n+1}\phi(x_{n+1}) - c\int dx_n\phi(x_n) = a$$

may be reduced to

$$\int \phi(x_n)\, dx_n = \frac{1-c^n}{1-c}a + c^n b.$$

The next artifice, also attributed to Laplace, is a method for finding the coefficients of a power series, and is very similar to that employed by Babbage in his paper, 'On some new methods of investigating the sums of several classes of infinite series', published in the *Philosophical Transactions*, 1819, **109**, 249–82.

Suppose $f(v) = A + A_1 v + A_2 v^2 + \cdots$ and it is required to find A_n. Then

$$v^{-n} f(v) = \frac{A}{v^n} + \frac{A_1}{v^{n-1}} + \cdots + A_n + A_{n+1} v + \cdots$$

and substituting $1/v$ for v,

$$v^n f(1/v) = A v^n + A_1 v^{n-1} + \cdots + A_n + A_{n+1}(1/v) + \cdots$$

Now putting

$$v + (1/v) = 2 \cos \theta,\ v^n + v^{-n} = 2 \cos n\theta,$$

$$\tfrac{1}{2}[v^{-n} f(v) + v^n f(1/v)] = A \cos n\theta + A_1 \cos \overline{n-1}\theta + \cdots + A_n$$
$$+ A_{n+1} \cos \theta + \cdots$$

Finally, multiplying by $d\theta$, integrating from $\theta = 0$ to $\theta = \pi$ we obtain

$$A_n = \frac{1}{2\pi} \int d\theta [v^{-n} f(v) + v^n f(1/v)]$$

(*ibid.*, p. 117).

Apart from the uncritical manipulation of infinite series and complex numbers, this does seem a very complicated way of obtaining a simple result. At this point Babbage makes one of his very few concessions to the question of convergence, for he asserts (*ibid.*, p. 119) that in the equation $f(\theta) = A + B \cos \theta + C \cos 2\theta + \cdots$ on integrating between w and $\pi + w$, where w is very small, we obtain

$$\int f(\theta)\, d\theta = A\pi - 2w(B + C + D + \cdots)$$

so that care must be taken to ensure that $(B + C + D + \cdots)$ is not infinite.

The last artifice of any interest in the chapter is concerned with the summation of series and arose from a correspondence between Goldbach and Euler (*ibid.*, pp. 121–3).

It is required to sum a series of the form

$$1 + \frac{1}{2^m}\left(1 + \frac{1}{2^n}\right) + \frac{1}{3^m}\left(1 + \frac{1}{2^n} + \frac{1}{3^n}\right) + \cdots$$

The technique is to consider the more general series

$$z = \frac{x}{1^m}\frac{y}{1^n} + \frac{x^2}{2^m}\left(\frac{y}{1^n} + \frac{y^2}{2^n}\right) + \frac{x^3}{3^m}\left(\frac{y}{1^n} + \frac{y^2}{2^n} + \frac{y^3}{3^n}\right) + \cdots$$

Then replacing x and y by e^x and e^y, respectively,

$$z = \frac{e^x}{1^m}\left(\frac{e^y}{1^n}\right) + \frac{e^{2x}}{2^m}\left(\frac{e^y}{1^n} + \frac{e^{2y}}{2^n}\right) + \cdots$$

Differentiating this m times with respect to x and n times with respect to y, we obtain

$$\frac{d^{m+n}z}{dx^m\,dy^n} = e^x e^y + e^{2x}(e^y + e^{2y}) + e^{3x}(e^y + e^{2y} + e^{3y}) + \cdots$$

Multiplying by e^x,

$$e^x\frac{d^{m+n}z}{dx^m\,dy^n} = e^{2x}e^y + e^{3x}(e^y + e^{2y}) + \cdots,$$

and subtracting

$$(1 - e^x)\frac{d^{m+n}z}{dx^m\,dy^n} = e^{x+y} + e^{2x+2y} + \cdots = \frac{e^{x+y}}{1 - e^{x+y}},$$

so that finally, on integration,

$$z = \int^{m+n} \frac{e^{x+y}\,dx^m\,dy^n}{(1 - e^x)(1 - e^{x+y})}.$$

When the result of this has been obtained, put $x = 0$, $y = 0$, and the required sum of the series remains.

If this chapter has demonstrated little original work, and if the mathematics with a few exceptions has not been as exciting as we might have been led to believe by Babbage's enthusiasm, it has at least indicated the source of several of his ideas.

The last chapter (*ibid.*, pp. 124–43), entitled 'Of questions requiring the invention of new modes of analysis', is both difficult

and original. Babbage attempts to give a mathematical formulation to some logical problems which normally defy such an
analysis. These problems can be put mostly in the category of
party games or tricks, but require great subtlety of argument to
express in mathematical terms.

He begins by making the point that problems which arise from
physical science usually have a neat algebraic formulation which
produces a problem that can either be solved directly or by
approximation. He proposes to deal with a different type of
logical problem:

> The nature of these questions to which we shall now direct our
> attention is entirely different and for by far the greater part of
> them all known methods are inefficient: the first great difficulty
> then presented to us is that of representing in symbolic lan-
> guage the conditions of the problem: unless this can be accom-
> plished all hope of solution must be given up and an approxima-
> tion supposing the question to admit of one cannot be disco-
> vered. The class of questions to which I allude chiefly comprise
> such as are referable to the Geometry of Situation and have very
> frequently arisen from games of skill. (*Ibid.*, p. 125.)

The problems are certainly difficult. On the next page he alludes
to a remark by Leibniz that 'Few occasions call forth the ingenuity
of mankind more than those games which they contrive for the
occupation of their leisure.'

He distinguishes between three types of game: those wholly
dependent on skill, those wholly dependent on chance and those
in which skill and chance both play a part. These can be distinguished further into games for one person and games played by
more than one. In individual games, like solitaire, the solution
may always be seen by retracing the steps that led to the goal.
Games of pure chance can usually be reduced to a series of
algebraic operations, but those of skill are much harder to
analyse.

The first difficulty to be faced is to produce an algebra which
corresponds to a geometry of relative position rather than distance. Geometry was almost completely concerned with metrical
relations before Babbage's time and algebraic geometry was a
mapping of these functions of distance onto an analytical
framework. Babbage seems unaware of the researches in projec-
tive geometry by Monge and Poncelet, but, even if he had been,
there would have been no suitable algebraic translation of results

available. In the theory of games, distance is almost always irrelevant, and properties of relative position fundamental, so it was necessary to invent his own algebraic interpretation.

The most important example here is an analysis of the game of noughts-and-crosses. After a surprisingly lengthy explanation of the rules, he attempts a mathematical formulation. The basic problem is one that appears not to have been previously considered in the history of mathematics: 'The first object is to discover some method of expressing any one indifferently out of n quantities or positions and when this is abstracted to express any one indifferently out of the remaining $n-1$' (*ibid.*, p. 137).

If the objects chosen in order are $x, x_1, x_2, \ldots$, Babbage uses the notation $f(x)$ to represent any of the nine cells, $f(x, x_1)$ for any of the remaining eight after x has been selected, $f(x, x_1, x_2)$ for the seven after x and x_1 have been taken, etc. The problem is to find a suitable formulation for the functions $f(x), f(x, x_1), f(x, x_1, x_2), \ldots$

He uses the convenient device of the Latin square with rows, columns and diagonals adding up to fifteen:

$$
\begin{array}{ccc}
4 & 9 & 2 \\
3 & 5 & 7 \\
8 & 1 & 6
\end{array}
$$

Then the variables $x, x_1, x_2, \ldots$ become numbers from 1 to 9 and the game becomes isomorphic with two players A and B alternately selecting numbers between 1 and 9, the object being to find any three which add up to 15.

Bearing this aim in mind until later, Babbage now introduces a refinement of one of his favourite techniques concerning the complex roots of unity.

Let S_x^p denote the sum of the xth powers of the pth roots of unity divided by p. Then, as usual, if x is a multiple of p, $S_x^p = 1$, if x is not a multiple of p, $S_x^p = 0$.

Also, and here is the refinement essential for the subsequent calculation, let S_n^p denote indifferently any one of the quantities $S_1^p, S_2^p, \ldots S_{p-1}^p$. It would appear at first sight that all of these quantities are zero, which cannot be of much use. Babbage obviously does not intend this, and we must modify his definition of S in view of the next statement (*ibid.*, p. 139), which indicates his meaning more clearly.

> If $p = 3$, S_x^3 will denote any one of the quantities $S_v^3, S_{v+1}^3, S_{v+2}^3$, according to the form of x: and the two which remain will be

S_{x+1}^3 and S_{x+2}^3; out of these we wish to select one indifferently which will depend on a new variable x_1, and as there are only two values these must be connected with the function $S_{x_1}^2$; the object will be accomplished if the x in S_x^3 is increased either by one or by 2 which will be the case if we take $S_{x+S_{x_1}+2S_{x_1}+1}^3$ and a similar mode of reasoning in the case of $p = 4$ will give $S_x^4, S_{x+S_{x_1}^3+2S_{x_1}^3+1+3S_{x_2}^3+2}^4$.

This sentence becomes meaningful only if we change the definition of S_n^p as any one of $S_1^p, S_2^p, \ldots S_{p-1}^p$ to being any one of $S_1^p, S_2^p, \ldots S_{p-1}^p, S_p^p$ so that it may be any one of p quantities, one of which has the value of one and the rest zero. It is apparent from the definition of S_x^p that $S_{x+np}^p = S_x^p$, where n is any positive integer.

From the above quotation it seems that for the noughts and crosses problem, $f(x) = S_x^9$,

$$f(x, x_1) = S_{x+2S_{x_1}^8+1+3S_{x_1}^8+2+4S_{x_1}^8+3+5S_{x_1}^8+4+6S_{x_1}^8+5+7S_{x_1}^8+6+8S_{x_1}^8+7}^9$$

with a very complicated expression for $f(x, x_1, x_2)$!

Realising this, Babbage tries a different stratagem with this notation: 'S_x^p will denote any of the quantities $S_1^p, S_2^p, \ldots S_p^p$ the remaining ones may be found by putting $x + 1, x + 2, \ldots x + p - 1$ successively for x in S_x^p and they are $S_{x+1}^p \, S_{x+2}^p \, S_{x+3}^p \, S_{x+p-1}^p$ (a) ' (*ibid.*, p. 140). This bears out the proper definition of S_x^p mentioned above. He continues on the same page, 'If we take S_x^{p-1}, one of the quantities $S_x^{p-1}, S_{x+1}^{p-1}, \ldots S_{x+p-1}^{p-1}$ will be equal to unity and all the rest zero consequently any one of the quantities (a) will be expressed by

$$S_{x+1}^p \, S_{x_1}^{p-1} + S_{x+2}^p \, S_{x_1+1}^{p-1} + \cdots S_{x+p-1}^p \, S_{x_1+p-2}^{p-1}, \text{'}$$

The other quantities in (a) are obtained by replacing x in this latter expression in turn by $x + 1, x + 2, \ldots x + p - 2$. Then if each of these quantities is similarly multiplied respectively by $S_{x_2}^{p-2}, S_{x_2+1}^{p-2}, \ldots S_{x_2+p-3}^{p-2}$, and added, the sum of these products will represent indifferently any of the $p - 2$ quantities which remain. In this way, we now have expressions (still quite complicated) for $f(x)$, $f(x, x_1)$, $f(x, x_1, x_2)$ and a systematic procedure for writing down the remaining functions. Thus the required technique for obtaining the successive functions has been evolved, and it remains only to apply it to the Latin square version of the noughts-and-crosses game.

This is relatively straightforward now, as Babbage explains. We first have to select indifferently any of the quantities 4, 9, 2, 3, 5, 7, 8, 1, 6, then any of the remaining eight, then any of the remaining seven, etc. All that has to be done is to substitute for S_x^p in the above expression the quantity

$$4S_x^9 + 9S_{x+1}^9 + 2S_{x+2}^9 + 3S_{x+3}^9 + 5S_{x+4}^9 + 7S_{x+5}^9 + 8S_{x+6}^9 + 1S_{x+7}^9 + 6S_{x+8}^9$$

which always reduces itself to one term, on account of the nature of the expression S_x^9.

He ends this remarkable analysis by noticing that in deriving each successive function the number of terms is multiplied by $p - 1$, where S_x^p refers to the previous function. This means that to analyse the whole game requires $9 + 9 \cdot 8 + 9 \cdot 8 \cdot 7 + \cdots + 9 \cdot 8 \cdot 7 \cdots 2 \cdot 1$ terms. He remarks 'The plan was proposed rather with a view of showing the possibility of the thing than as being convenient for executing it' (ibid., p. 141).

It is still without doubt a considerable achievement to have analysed such a novel sort of problem in this successful way, and even if we have not learnt anything new about the strategy to adopt in playing this game, or how to programme a computer to do the work for us, at least we have a calculus to apply to this type of question. This analysis of a chain of random choices, each dependent upon the previous one, and working towards a goal, must count as the first recorded stochastic process in the history of mathematics.

To summarise this chapter, we have here a very mixed display of the author's strengths and weaknesses. The whole work is in a very unpolished form, suggesting that Babbage probably never gave it a second look. It is likely that at this time he became so involved in the construction of his calculating engines, that he virtually retired from research in pure mathematics. The mathematical world is the poorer through Babbage never having developed nor published the 'Philosophy of Analysis', since, as has been shown, the work contains some exceptionally fine material with the promise of better to come. It is almost tragic to think that the author spent most of his remaining fifty years trying to devise suitable machinery for his engines. If he had developed the very fruitful ideas contained in the book, and eliminated much of the dross before publishing, then it might well have been that mathematical philosophy, modern algebra, the theory of games and stochastic mathematics would have developed many decades before they actually did.

NOTES

1. British Museum Additional Manuscripts 37182, No. 323.
2. *Ibid.*, No. 371.
3. *Ibid.*, No. 415.
4. *Ibid.*, No. 411.
5. *Ibid.*, No. 105.
6. *Ibid.*, No. 407.
7. *Edinburgh Encyclopedia*, 1830, **15**, 396.
8. G. Peacock, *A Treatise on Algebra*, Cambridge, 1830, p. 1.
9. *Ibid.*, p. vi.
10. *Ibid.*, p. vii.
11. *Ibid.*, p. viii.
12. *Ibid.*, p. xiv.
13. *Ibid.*, pp. xvi–xvii.
14. *Ibid.*, pp. xvi–xvii.
15. *Ibid.*, pp. xviii–xix.
16. British Museum Additional Manuscripts 37187, No. 563.
17. British Museum Additional Manuscripts 37194, No. 375.
18. F. Cajori, *A History of Mathematics*, New York, 1919, second edition, p. 331.
19. *Philosophical Transactions*, 1817, **100**, 197–216.

6

Miscellaneous papers in analysis, probability and geometry

I include in this chapter a description of some of Babbage's early mathematical papers which appear isolated from the general subjects considered elsewhere. All show in different ways the fruitful imagination of the author and, although the mathematical standard is not consistently high, they are still valuable in illustrating his processes of creativity.

The first paper, written when Babbage was an undergraduate at Cambridge, is taken from the *Memoirs of the Analytical Society*, Cambridge, 1813, pp. 1–31. According to Babbage this book was written solely by J. F. W. Herschel and himself, his own contribution consisting of the preface and the first chapter, 'On continued products', which we will now consider.

This work contains some interesting results on expansions of functions as finite products. Babbage is careful not to say too much about infinite products, which is fortunate because, as we shall see in other papers, like most of his contemporaries he fell into serious error when considering infinite series and products.

Before this time plenty of work both correct and incorrect, had been done concerning finite and infinite series. There seems, however, to have been little interest in products apart from Vieta's demonstration that

$$\frac{2}{\pi} = \sqrt{\frac{1}{2}} \cdot \sqrt{\left(\frac{1}{2} + \frac{1}{2}\sqrt{\frac{1}{2}}\right)} \cdot \sqrt{\left[\frac{1}{2} + \frac{1}{2}\sqrt{\left(\frac{1}{2} + \frac{1}{2}\sqrt{\frac{1}{2}}\right)} \cdots \right]}$$

and Wallis's interpolation to show that

$$\frac{4}{\pi} = \frac{3 \cdot 3 \cdot 5 \cdot 5 \cdot 7 \cdot 7 \cdots}{2 \cdot 4 \cdot 4 \cdot 6 \cdot 6 \cdot 8 \cdots}.$$

Babbage does quote one result from Euler on finite products, but it seems that this paper was one of the first to treat this subject systematically.

He first proves some general theorems in the calculus of functions, which are used later in deriving products.

He considers first (*ibid.*, p. 1) the equation

$$\psi x \cdot \phi x = \chi x. \tag{a}$$

Then if a function f can be found so that $\phi \mathrm{f} x = \chi x$, the equation becomes

$$\psi x \cdot \phi x = \phi \mathrm{f} x. \tag{b}$$

x is now replaced successively by $\mathrm{f} x, \mathrm{f}^2 x, \ldots \mathrm{f}^n x$ giving the equations

$$\psi \mathrm{f} x \cdot \phi \mathrm{f} x = \phi \mathrm{f}^2 x,$$

$$\psi \mathrm{f}^2 x \cdot \phi \mathrm{f}^2 x = \phi \mathrm{f}^3 x,$$

$$\vdots$$

$$\psi \mathrm{f}^n x \cdot \phi \mathrm{f}^n x = \phi \mathrm{f}^{n+1} x,$$

and multiplying all the equations together and cancelling (*ibid.*, p. 1),

$$\phi \mathrm{f}^{n+1} x / \phi \mathrm{f} x = \psi \mathrm{f} x \cdot \psi \mathrm{f}^2 x \cdots \psi \mathrm{f}^n x \cdots \tag{c}$$

This equation, as we shall see, is useful in the subsequent theory for the finite product on the right. Provided that from (*a*) relative expressions for the functions f, ϕ and ψ can be found, (*c*) can be used to supply an appropriate product formula, which can in some circumstances be extended into an infinite product.

Returning to equation (*b*), ψx is now replaced by $(\psi x)^{p-1}$, giving

$$(\psi x)^{p-1} \phi x = \phi \mathrm{f} x \cdots \tag{d}$$

and on division by $(\psi x)^p$,

$$\frac{\phi x}{\psi x} = \frac{\phi \mathrm{f} x}{(\psi x)^p} \cdots \tag{e}$$

He now introduces

$$y = \left\{ \frac{^{1/p}\phi \mathrm{f} x}{\psi \mathrm{f} x} \right\{ \frac{^{1/p}\phi \mathrm{f}^2 x}{\psi \mathrm{f}^2 x} \right\{ \frac{^{1/p}\phi \mathrm{f}^3 x}{\psi \mathrm{f}^3 x} \right\{^{1/p} \cdots \left\{ \frac{^{1/p}\phi \mathrm{f}^n x}{\psi \mathrm{f}^n x} \right\},$$

where it is explained in a footnote (*ibid.*, p. 2) that the symbol $\{^{1/p}$ signifies that the $1/p$ th power is to be taken of all that follows the symbol. This means in fact that

$$y = (\phi fx/\psi fx)^{1/p} \cdot (\phi f^2 x/\psi f^2 x)^{1/p^2} \cdots (\phi f^n x/\psi f^n x)^{1/p^n} \cdots \qquad \text{(I)*}$$

Multiplying both sides by $(\phi f^{n+1} x)^{p^{1/(n+1)}}$, this gives

$$y(\phi f^{n+1} x)^{p^{1/(n+1)}} = \left\{ {}^{1/p} \phi fx \left\{ {}^{1/p} \frac{\phi f^2 x}{(\psi fx)^p} \left\{ {}^{1/p} \frac{\phi f^3 x}{(\psi f^2 x)^p} \right\}^{1/p} \cdots \right. \right.$$
$$\cdot \left\{ {}^{1/p} \frac{\phi f^n x}{(\psi f^{n-1} x)^p} \left\{ {}^{1/p} \frac{\phi f^{n+1} x}{(\psi f^n x)^p} \right\} \right. . \qquad \text{(II)}$$

Applying (*e*) $\phi x/\psi x = \phi fx/(\psi x)^p$, it follows by continually replacing x by fx, $f^2 x$, ... $f^n x$ that

$$\frac{\phi fx}{\psi fx} = \frac{\phi f^2 x}{(\psi fx)^p} \quad \text{and} \quad \frac{\phi f^2 x}{\psi f^2 x} = \frac{\phi f^3 x}{(\psi f^2 x)^p} \cdots,$$

from which the equation can now be simplified to

$$(\phi f^{n+1} x)^{p^{1/(n+1)}} y = \left\{ {}^{1/p} \phi fx \left\{ {}^{1/p} \frac{\phi fx}{\psi fx} \left\{ {}^{1/p} \frac{\phi f^2 x}{\psi f^2 x} \right\}^{1/p} \cdots \left\{ {}^{1/p} \frac{\phi f^n x}{\psi f^n x} \right\}, \right. \right. \right.$$
$$\text{(III)}$$

the one given in the text (*ibid.*, p. 2).

Also, the right-hand side is identical with $[y \cdot \phi fx]^{1/p}$ so that

$$(\phi f^{n+1} x)^{p^{1/(n+1)}} y = [y \phi fx]^{1/p}. \qquad \text{(IV)}$$

It should be mentioned here that since equation (*e*) has been used it must be assumed that p is a number such that $(\psi x)^{p-1} = \psi x$. This is trivially true for $p = 2$ and possibly for other values, depending on the type of function ψx.

Proceeding from (IV), raising both sides to the power p, $y^p (\phi f^{n+1} x)^{1/p^n} = y \phi fx$, and on division by y,

$$y^{p-1} = \frac{\phi fx}{(\phi f^{n+1} x)^{1/p^n}},$$

therefore

$$\frac{\phi fx}{(\phi f^{n+1} x)^{1/p^n}} = \left\{ {}^{1/p} \frac{\phi fx}{\psi fx} \left\{ {}^{1/p} \frac{\phi f^2 x}{\psi f^2 x} \right\}^{1/p} \cdots \left\{ {}^{1/p} \frac{\phi f^{nx}}{\psi f^n x} \right\} . \right. \right. \qquad (f)$$

---

* The Roman numbers on the equations are for reference purposes when Babbage has not given a letter or number.

This completes the theoretical part of the argument, which seems perfectly sound, and we now look at particular cases of these general results.

The notation used for sums and products is S^n and P^n, respectively. Babbage does not always make clear where the summation or production begins but generally uses the suffix n over the S or P to denote where it ends. For example,

$$P^n\{\psi f^n x\} = \psi f x \cdot \psi f^2 x \cdot \psi f^3 x \cdots \psi f^n x.$$

We will assume, unless otherwise stated, that the first term in the sum or product is when $n = 1$.

For his first example, Babbage takes $\psi x = 1 + x + \cdots + x^a$, and $fx = x^{a+1}$. Summing the series for ψx as a geometric progression, this means that the equation for the function ϕ is given by

$$\frac{x^{a+1} - 1}{x - 1}\phi x = \phi x^{a+1}.$$

It is of no particular interest to follow how he solves this by the method of finite differences. It is sufficient to say that he arrives at a solution $\phi x = x - 1$, which obviously satisfies the functional equation. In this type of problem it is enough to obtain particular solutions for the functions rather than something more general.

Having shown that $\psi x = 1 + x + \cdots + x^a$, $fx = x^{a+1}$, and $\phi x = x - 1$ are solutions of the functional equation $\psi x \cdot \phi x = \phi f x$, equation (c) can now be applied (*ibid.*, p. 4), which gives immediately the relation:

$$\frac{x^{\overline{a+1}^{n+1}} - 1}{x^{\overline{a+1}^n} - 1} = P^n(1 + x^{1 \cdot \overline{a+1}^n} + x^{2 \cdot \overline{a+1}^n} + \cdots + x^{a \cdot \overline{a+1}^n}).' \qquad (1)$$

If a is given special values, several interesting results may be derived. Taking first $a = 1$,

$$\frac{x^{2^{n+1}} - 1}{x^2 - 1} = P^n(1 + x^{2^n}) = (1 + x^2)(1 + x^4) \cdots (1 + x^{2^n}), \qquad (2)$$

a result which can be verified directly by multiplication, and taking $a = 2$,

$$\frac{x^{3^{n+1}} - 1}{x^3 - 1} = P^n(1 + x^{3^n} + x^{2 \cdot 3^n}) = (1 + x^3 + x^6) \cdots (1 + x^{3^n} + x^{2 \cdot 3^n}).$$

$$(3)$$

Returning to equation (1), Babbage now introduces complex numbers and makes the substitution $x + x^{-1} = 2 \cos \theta$ (*ibid.*, p. 6). This means that $x - x^{-1} = 2\sqrt{(-1)} \sin \theta$, and, using De Moivre's theorem, that $x^r + x^{-r} = 2 \cos r\theta$ and $x^r - x^{-r} = 2\sqrt{(-1)} \sin r\theta$.

His general equation is somewhat obscure and it will be more indicative of the value of this substitution by applying it to (2) rather than (1) so that we take $a = 1$ and start from

$$\frac{x^{2^{n+1}} - 1}{x^2 - 1} = P^n(1 + x^{2^n}). \tag{2}$$

Now

$$P^n(1 + x^{2^n}) = P^n(x^{2^{n-1}} + x^{-2^{n-1}})x^{2^{n-1}}$$

$$= x \cdot x^2 \cdot x^{2^2} \ldots x^{2^{n-1}} P^n(x^{2^{n-1}} + x^{-2^{n-1}})$$

$$= x^{2^n - 1} P^n(2 \cos 2^{n-1} \theta) = \frac{x^{2^{n+1}} - 1}{x^2 - 1},$$

therefore

$$P^n(2 \cos 2^{n-1}\theta) = \frac{x^{2^{n+1}} - 1}{x^2 - 1} \cdot \frac{x}{x^{2^n}} = \frac{x^{2^n} - x^{-2^n}}{x - x^{-1}} = \frac{\sin 2^n \theta}{\sin \theta}$$

replacing θ by 2θ, this gives the result

$$\frac{\sin 2^{n+1}\theta}{\sin 2\theta} = 2^n P^n(\cos 2^n \theta)$$

which is equivalent, as Babbage remarks (*ibid.*, p. 7), to Euler's expression

$$`\frac{1}{2^n} \frac{\sin 2^{n+1}\theta}{\sin 2\theta} = \cos 2\theta \cdot \cos 2^2\theta \ldots \cos 2^n\theta \ldots(1, 1)'.$$

Similarly, taking $a = 3$, the result

$$\frac{1}{2^n} \frac{\sin 4^{n+1}\theta}{\sin 4\theta} = P^n(\cos 1 \cdot 4^n\theta + \cos 3 \cdot 4^n\theta)$$

is obtained.

This earliest published paper demonstrates clearly his talent for manipulation. His encounter with infinite series, however, was not so happy. He brings his great powers of inventiveness to this subject but like many of his contemporaries, appeared insensitive to problems arising out of convergence, and accepted unquestioningly several absurd results. The paper on this topic, 'On some new methods of investigating the sums of several classes of infinite series', was published in *Philosophical Transactions*, 1819, **109**, 249–82. The paper was signed 25 March 1819, and read on 1 April 1819. The ideas expressed in this paper, however, were known to Babbage several years previously. 'The process[es] which it is the object of this paper to explain, were discovered several years since, but certain difficulties connected with the subject, which I was at that time unable to explain, and which were equally inexplicable to several of my friends, to whom I had communicated these methods, induced me to defer publishing them, until I could offer some satisfactory solution' (*ibid.*, p. 249). This remark refers chiefly to the second part of the paper which is entitled 'Method of expanding horizontally and summing vertically'. We will see when examining this part that the solution eventually arrived at was far from satisfactory. He continues: 'This method, which I employed about the year 1812, gave the values of a variety of series whose sums had not hitherto been known, most of which were apparently correct, but some of the consequences which followed were evidently erroneous' (*ibid.*, p. 249).

I look first at his most diverting work in the first half of the paper, a highly original attempt to sum series of the form

$$\frac{A_1 x}{(\sin\theta)^n} + \frac{A_2 x^2}{(\sin 2\theta)^n} + \frac{A_3 x^3}{(\sin 3\theta)^n} \&\text{c.}$$

where n and the $A_1, A_2, A_3, \ldots$ are constants. The work begins on p. 253.

First, v^{2z} is put for x and the function ψx defined by $\psi x = A_1 x + A_2 x^2 + \cdots$, so that

$$\psi v^{2z} = A_1 v^{2z} + A_2 v^{4z} + A_3 v^{6z} + \cdots$$

We now 'integrate' both sides of this equation using the result $\sum v^{2iz} = v^{2iz}/(v^2 - 1)$. This appears to mean that z is replaced in turn by $z + 1, z + 2, \ldots$ obtaining

$$\psi v^{2z} = A_1 v^{2z} + A_2 v^{4z} + A_3 v^{6z} + \cdots,$$

$$\psi v^{2(z+1)} = A_1 v^{2(z+1)} + A_2 v^{4(z+1)} + A_3 v^{6(z+1)} + \cdots,$$

$$\psi v^{2(z+2)} = A_1 v^{2(z+2)} + A_2 v^{4(z+2)} + A_3 v^{6(z+2)} + \cdots,$$

$$\vdots \qquad \vdots$$

Summing the infinite geometric progressions in each column,

$$\sum \psi v^{2z} = A_1 \frac{v^{2z}}{v^2-1} + A_2 \frac{v^{4z}}{v^4-1} + A_3 \frac{v^{6z}}{v^6-1} + \cdots$$

There appears to be a mistake in sign here as actually

$$\sum v^{2iz} = \frac{v^{2iz}}{1-v^2}.$$

Performing this 'integration' n times,

$$\sum^n \psi v^{2z} = A_1 \frac{v^{2z}}{(v^2-1)^n} + A_2 \frac{v^{4z}}{(v^4-1)^n} + A_3 \frac{v^{6z}}{(v^6-1)^n} + \cdots$$

Now introducing complex numbers, let $v = \cos\theta \pm \sqrt{(-1)}\sin\theta$. As usual,

$$v + (1/v) = 2\cos\theta, \ v - (1/v) = 2\sqrt{(-1)}\sin\theta,$$

$$v^n + (1/v^n) = 2\cos n\theta, \ v^n - (1/v^n) = 2\sqrt{(-1)}\sin n\theta.$$

Then the equation becomes

$$[2\sqrt{(-1)}]^n \sum^n \psi v^{2z} = A_1 \frac{v^{2z-n}}{(\sin\theta)^n} + A_2 \frac{v^{4z-2n}}{(\sin 2\theta)^n} + \cdots,$$

and putting $z + (n/2)$ for z:

$$(2\sqrt{-1})^n \sum^n \psi v^{2z+n} = A_1 \frac{v^{2z}}{(\sin\theta)^n} + A_2 \frac{v^{4z}}{(\sin 2\theta)^n} + \cdots$$

$$= A_1 \frac{x}{(\sin\theta)^n} + A_2 \frac{x^2}{(\sin 2\theta)^n} + \cdots, \quad (1)$$

the expression that is wanted, provided that the left-hand side can be evaluated.

This is achieved by saying that if after the nth integration, v is replaced by $1/v$, the equation (1) becomes

$$[-2\sqrt{(-1)}]^n \sum^n \psi v^{-2z-n} = A_1 \frac{x^{-1}}{(\sin\theta)^n} + A_2 \frac{x^{-2}}{(\sin 2\theta)^n} + \cdots \quad (2)$$

Babbage now remarks that in normal circumstances neither (1) nor (2) can be evaluated but in certain simple cases this can be achieved by a combination of the two results.

Suppose, for example, that $A_1 = 1$, $A_2 = -1$, $A_3 = 1$, $A_4 = -1$, &c. Then

$$\psi v^{2z+n} = v^{2z+n} - v^{2(2z+n)} + v^{3(2z+n)} - \cdots = \frac{v^{2z+n}}{1 + v^{2z+n}}$$

and

$$\psi v^{-2z-n} = \frac{v^{-2z-n}}{1 + v^{-2z-n}}.$$

Letting

$$n = 1, \; 2\sqrt{(-1)} \sum \left(\frac{v^{2z+1}}{1 + v^{2z+1}} \right)$$

$$= A_1 \frac{x}{\sin \theta} + A_2 \frac{x^2}{\sin 2\theta} + \cdots,$$

and

$$-2\sqrt{(-1)} \sum \frac{v^{-2z-1}}{1 + v^{-2z-1}} = A_1 \frac{x^{-1}}{\sin \theta} + A_2 \frac{x^{-2}}{\sin 2\theta} + \cdots$$

On subtraction,

$$2\sqrt{(-1)} \sum \left(\frac{v^{2z+1}}{1 + v^{2z+1}} + \frac{v^{-2z-1}}{1 + v^{-2z-1}} \right)$$

$$= A_1 \frac{x - x^{-1}}{\sin \theta} + A_2 \frac{x^2 - x^{-2}}{\sin 2\theta} + \cdots$$

Now

$$\frac{v^{2z+1}}{1 + v^{2z+1}} + \frac{v^{-2z-1}}{1 + v^{-2z-1}} = \frac{v^{2z+1}}{1 + v^{2z+1}} + \frac{1}{v^{2z+1} + 1} = 1,$$

and the problem becomes how to deal with the expression $\sum (1)$.

Babbage now spoils his calculation by asserting that $\sum (1) = z + b$, where b is an arbitrary constant. There can surely be no justification for this. The result can only be obtained on the assumption that the summation implies an integration in the normal sense of the word, but $\sum$ has clearly not been used in this way earlier in the calculation. Instead, Babbage appears to have interpreted $f(z)$ as $f(z) + f(z + 1) + f(z + 2) + \cdots$ and to be consistent it would seem that $\sum (1) = 1 + 1 + 1 + \cdots$, a result hardly likely to lead to any useful consequences. By taking $\sum (1) = z + b$ and manoeuvring the series legitimately, Babbage arrived at his first conclusion,

$$\frac{\theta'}{\theta} = \frac{\sin \theta'}{\sin \theta} - \frac{\sin 2\theta'}{\sin 2\theta} + \frac{\sin 3\theta'}{\sin 3\theta} - \ldots$$

We can only remark that this expression is true for $\theta' = 0$, certainly untrue for θ' as any multiple of π, untrue even for $\theta = \theta'$ assuming one of Euler's dubious results for infinite series, and that in any case the series is hopelessly non-convergent for all θ and θ' except in the trivial case when $\theta' = 0$, $\theta \neq 0$. As the rest of the results in the first part of this paper depend on the same erroneous principle there is no point in enumerating them any further.

The second part is a sad example of the indifference of a mathematician of Babbage's calibre to the most elementary feelings for convergence. He begins (*ibid.*, p. 268) with a bold assertion of four wrong results,

$$`0 = 1^{2n} - 2^{2n} + 3^{2n} - 4^{2n} + \&c.$$

$$\tfrac{1}{2} = \cos \theta - \cos 2\theta + \cos 3\theta - \&c.$$

$$\tfrac{1}{2} = \cos \theta + \cos 2\theta + \cos 3\theta + \&c.$$

$$0 = 1^{2n+1} - 3^{2n+1} + 5^{2n+1} - \&c.,'$$

and he actually says, 'It is unnecessary to give proofs of these and other similar ones which have been frequently noticed, as they may be very easily demonstrated.' The remainder of the paper is based on these results and therefore has very little interest.

Babbage's only published work on the theory of numbers, one of the more demanding branch of mathematics as far as rigour of proof is concerned, was slightly more successful. The title is 'Demonstration of a theorem relating to prime numbers' and it appeared in the *Edinburgh Philosophical Journal*, 1819, **1**, 46–9.

The work is concerned with an extension of Wilson's theorem, that the number $(n-1)! + 1$ is divisible by n if n is a prime number, otherwise it is not divisible by n. This theorem first appeared in Waring's *Meditationes Analyticae*, Cambridge, 1776, but neither Waring nor Wilson could supply a proof. It had then aroused the interest of Continental analysts, and Lagrange and Euler had given several proofs. Babbage's aim was to provide a function of the integer n which had a similar property with respect to n^2. His conclusion was: 'The theorem which I have arrived at is as follows, that

$$\frac{n+1 \cdot n+2 \cdot n+3 \ldots \overline{2n-1}}{1 \cdot 2 \cdot 3 \ldots n-1} - 1$$

is always divisible by n^2 when n is a prime number, otherwise it is not.' (*Edinburgh Philosophical Journal*, **1**, p. 46.) He adds later that $n = 2$ is, of course, an exception to the rule.

His proof is a simple one. Starting from the result that the sum of the squares of the coefficients in a binomial series with index n is equal to the coefficient of the middle term of the binomial series with index $2n$, he writes down the identity

$$\frac{2n \cdot 2n-1 \cdot 2n-2 \cdot \ldots n+1}{1 \cdot 2 \cdot 3 \cdot \ldots n}$$
$$= 1^2 + \binom{n}{1}^2 + \binom{n}{2}^2 + \cdots + \binom{n}{n-1}^2 + 1^2.$$

Subtracting 2 from each side and simplifying,

$$2\left(\frac{n+1 \cdot n+2 \cdot \ldots 2n-1}{1 \cdot 2 \cdot 3 \cdot \ldots n-1} - 1\right) = \binom{n}{1}^2 + \binom{n}{2}^2 + \cdots + \binom{n}{n-1}^2.$$

But $\binom{n}{1}, \binom{n}{2}, \ldots$ are all divisible by n when n is prime and hence the sum of their squares is divisible by n^2. It can be seen at this stage that, since the factor 2 appears on the left, the theorem is unlikely to be true for $n = 2$.

The proof is not complete, for Babbage says nothing about the 'otherwise it is not' part of the theorem. It can be seen by trying for example $n = 4, 6, \ldots$ that the function of n is then not divisible by n^2 but there is no guarantee that this negative assertion is general. It would have been better if Babbage had omitted the words 'otherwise it is not' from his formulation of the theorem, for then his proof is certainly valid.

He next supplies a generalisation of the result, making use of Euler's identity that if p, n, q are positive integers, then

$$\binom{p+n}{q+n} = \binom{n}{0}\binom{p}{q} + \binom{n}{1}\binom{p}{q+1} + \binom{n}{2}\binom{p}{q+2} + \cdots$$

Letting $q = p - n$, it follows that

$$\binom{p+n}{p} = \binom{n}{0}\binom{p}{p-n} + \binom{n}{1}\binom{p}{p-n+1} + \cdots + \binom{n}{n}\binom{p}{p}$$

and all terms on the right, except the first and last, are divisible by both n and p if both numbers are prime. Consequently,

$$\frac{p+n \cdot p+n-1 \cdot \ldots n+1}{1 \cdot 2 \cdot 3 \cdot \ldots p} - \frac{p \cdot p-1 \cdot \ldots n+1}{1 \cdot 2 \cdot \ldots p-n} - 1$$

is always divisible by np when n and p are prime. In the case when $n = p$ the result reduces to the theorem proved by Babbage.

On the other hand, if p is prime and $p > n$, then since

$$\frac{p \cdot p-1 \cdot \ldots n+1}{1 \cdot 2 \cdot \ldots p-n}$$

is divisible by p, it follows that

$$\frac{p+n \cdot p+n-1 \cdot \ldots n+1}{1 \cdot 2 \cdot 3 \cdot \ldots p} - 1$$

is always divisible by p whatever n is, provided that p is prime.

It is unfortunate that this short but competent paper ends with an error, for Babbage asserts that this expression is also divisible by n, when clearly it is not.

The next paper is Babbage's only one on probability. Babbage showed a lifelong interest in problems of statistics and their relation to theories of economics and insurance calculations, but in this paper his concern was primarily mathematical, and we will find considerable skill in the manipulation of functions and polynomials.

The title is 'An Examination of some questions connected with games of chance', published in the *Transactions of the Royal Society of Edinburgh*, 1823, 9, 153–77, and actually read before the Society on 21 March 1820. The paper takes the form of a series of problems arising from gambling games in which the emphasis is on mathematical generalisation.

The final problem is a good illustration of Babbage's power of algebraic manipulation. A player draws a number of balls in succession numbered 1, 2, 3, ... k from a bag which contains many of each kind. With each ball is associated a number $n_1, n_2, \ldots n_k$, so that if a number 1, 2, ... k is drawn, then the corresponding profit is respectively $n_1, n_2, \ldots$ or n_k times the amount staked. The initial stake is u. If the last ball to be drawn is numbered i, then vn_i is added to the previous total stake before the next one is drawn. The problem is to calculate the total profit after a game in which ball-1 has been drawn p times, ball-2 q times, ball-3 r times, and so on.

The solution is quite brief and elegant. A number is required which can take at random the value n_1, or n_2, ... or n_k. To achieve this Babbage employs a device used also in his solution to the game of noughts-and-crosses (see chapter 5). Let $\alpha, \beta, \gamma, \ldots$ be the kth roots of unity. Then

$$S_a = \frac{\alpha^a + \beta^a + \gamma^a + \cdots}{k}$$

takes the value 0 except when a is a multiple of k in which case its value is 1. Consequently, if

$$P_a = n_1 S_a + n_2 S_{a+1} + \cdots + n_k S_{a+k-1},$$

one and only one of the numbers $a, a+1, \ldots a+k-1$ will be a multiple of k with the result that $P_a = n_1$ or n_2 or ... or n_k, depending on the value of a.

Applying this to the gaming problem, the successive profits will be

$$u P_a,$$

$$u P_b + (v P_a) P_b,$$

$$u P_c + v (P_a + P_b) P_c, \text{ etc.}$$

This means that the coefficient of u will be the second term, and the coefficient of v the third term of the polynomial expansion $(x + P_a)(x + P_b) \ldots$ But it is given that, of the quantities $P_a, P_b, \ldots, p$ are equal to n_1, q to $n_2, \ldots$ so that the polynomial becomes $(x + n_1)^p (x + n_2)^q + \cdots$ from which the appropriate coefficients may be easily derived.

The major interest in this particular paper has been the ingenious use of algebra to solve probability problems, especially the devices using the coefficients of polynomials and the roots of unity. It would seem perfectly fair to agree with the author in his conclusion that

> The questions examined in the preceding pages afford an instance of the immediate application of some very abstract propositions of analysis to a subject of constant occurrence, which being as far as I have been able to discover, hitherto untouched, and also requiring reasoning of rather an unusual nature, I have preferred treating in particular instances,

instead of investigating it in its most general form. (*Ibid.*, p. 177.)

In the course of his mathematical career, Babbage published three papers of a geometrical nature. Geometry does not seem to have been one of his major interests and the papers are dispersed over the years 1816, 1820 and 1827, with no suggestion in his other work that he thought much about this subject between these times.

The point has already been made that, in British mathematics generally at this time, geometers appear to have been as unaffected by Descartes's work as analysts were by Leibniz's. Their books and papers consisted invariably of Euclidean geometry performed by the normal synthetic methods. Babbage was one slight exception to the rule. His work is almost entirely algebraic or trigonometrical, and he usually solves his problem, or rather chooses it in such a way that no classical subtleties are required in the argument. His greatest skill is in the manipulation of algebraic equations and his attitude to geometry appears to be to use it to provide illustrations of analytical procedures rather than for any intrinsic merit of its own.

In the second and third papers Babbage attempts to join the line of mathematicians who have tried to restore Euclid's lost work on porisms. The first paper deals with the ordinary geometry of the circle. It too was historically inspired, as it attempts to prove some theorems stated seventy years earlier. The full title is 'Demonstration of some of Dr. Matthew Stewart's general theorems, to which is added, an account of some new properties of the circle', and was published in the *Journal of Sciences and the Arts*, London, 1817, 1, 6–24.

Matthew Stewart's work was entitled *Some General Theorems of considerable use in the higher parts of Mathematics*, and was published in 1746. He was then Minister at Rosneath and one of Scotland's leading mathematicians. His book contains sixty-four geometrical propositions of a highly complicated nature but with very few proofs, and justifies this in the preface by stating:

> The theorems contained in the following sheets are given without being demonstrated, excepting the first five: and as they are entirely new, save one or two at most, the author expects their being published even in this way may be agreeable to those that are not unaccustomed to speculations of this kind. Such will easily allow, that to explain, in a proper way, so many

theorems, so general, and of so great difficulty as most of these are, would require a greater expense of time and thought than can be expected soon from one in the author's situation. He therefore thought it was better they should appear in the way they now are, than lie by him till an uncertain hereafter. If any give themselves the trouble to explain some of these theorems, they will find their time and pains sufficiently rewarded, by the discovery of several new and curious propositions that otherwise might have escaped their observations.[1]

We are left to guess whether Stewart was able to give the proofs of some of these theorems of a very general nature. It is possible that he obtained correct results by unjustifiable inductions from simpler cases. At any rate, no one seems to have supplied the missing proofs, for as Babbage remarks on p. 6 of his own paper in the *Journal of Sciences and the Arts*, 'The means by which Dr. Stewart arrived at these theorems he never made public, nor does it appear that any traces of them have been found among his papers.'

Babbage commences his work with some lemmas concerning trigonometric series. The first one is a well-known, easily proved result about the addition of cosines and sines of angles in arithmetic progression: that

$$\cos \theta + \cos (\theta + b) + \cos (\theta + 2b) + \cdots + \cos (\theta + \overline{p-1}b)$$

$$= \frac{\cos \left(\theta + \frac{b}{2} \right) \sin \frac{p}{2} b}{\sin \frac{b}{2}}$$

and

$$\sin \theta + \sin (\theta + b) + \sin (\theta + 2b) + \cdots + \sin (\theta + \overline{p-1}b)$$

$$= \frac{\sin \left(\theta + \frac{b}{2} \right) \sin \frac{p}{2} b}{\sin \frac{b}{2}}.$$

Now if $b = 2\pi/p$, then the right-hand side in both the above expressions becomes zero so that

$$\cos \theta + \cos \left(\theta + \frac{2\pi}{p} \right) + \cdots + \cos \left(\theta + \overline{p-1}\frac{2\pi}{p} \right) = 0$$

and

$$\sin \theta + \sin \left(\theta + \frac{2\pi}{p}\right) + \cdots + \sin \left(\theta + \overline{p-1}\frac{2\pi}{p}\right) = 0.$$

Babbage now adds his more original lemma 2 (*ibid.*, p. 8) that, if q is any odd number,

$$[\cos \theta]^q + \left[\cos \left(\theta + \frac{2\pi}{p}\right)\right]^q + \cdots + \left[\cos \left(\theta + \overline{p-1}\frac{2\pi}{p}\right)\right]^q = 0$$

and

$$[\sin \theta]^q + \left[\sin \left(\theta + \frac{2\pi}{p}\right)\right]^q + \cdots + \left[\sin \left(\theta + \overline{p-1}\frac{2\pi}{p}\right)\right]^q = 0.$$

This is proved by observing that if q is odd, then $\cos^q \phi$ may be expressed, using De Moivre's theorem, for example, as

$$\cos^q \phi = A \cos \phi + B \cos 3\phi + \cdots + L \cos q\phi$$

and similarly

$$\sin^q \phi = A' \sin \phi + B' \sin 3\phi + \cdots + L' \sin q\phi,$$

where $A, B, \ldots L, A', B', \ldots L'$ are all constants. Then the sum

$$[\cos \theta]^q + \left[\cos \left(\theta + \frac{2\pi}{p}\right)\right]^q + \cdots + \left[\cos \left(\theta + \overline{p-1}\frac{2\pi}{p}\right)\right]^q$$

can be expressed as:

$$A \cos \theta + B \cos 3\theta + \cdots + L \cos q\theta$$

$$+ A \cos \left(\theta + \frac{2\pi}{p}\right) + B \cos 3 \left(\theta + \frac{2\pi}{p}\right) + \cdots + L \cos q \left(\theta + \frac{2\pi}{p}\right)$$

$$\vdots$$

$$+ A \cos \left(\theta + \overline{p-1}\frac{2\pi}{p}\right) + B \cos 3 \left(\theta + \overline{p-1}\frac{2\pi}{p}\right) + \cdots$$

$$+ L \cos q \left(\theta + \overline{p-1}\frac{2\pi}{p}\right).$$

Then using the first lemma, the sum of each column is zero, so that the total sum of the series is also zero. Similarly, the result for sines can be proved.

The third lemma (p. 8) is that if q is any even number, then

$$[\cos\theta]^q + \left[\cos\left(\theta + \frac{2\pi}{p}\right)\right]^q + \cdots + \left[\cos\left(\theta + \overline{p-1}\frac{2\pi}{p}\right)\right]^q$$

$$= \frac{p}{2^q}\frac{q \cdot q - 1 \ldots \left(\frac{q}{2}+1\right)}{1 \cdot 2 \ldots \frac{q}{2}}$$

and

$$[\sin\theta]^q + \left[\sin\left(\theta + \frac{2\pi}{p}\right)\right]^q + \cdots + \left[\sin\left(\theta + \overline{p-1}\frac{2\pi}{p}\right)\right]^q$$

$$= \frac{p}{2^q}\frac{q \cdot q - 1 \cdot \ldots \left(\frac{q}{2}+1\right)}{1 \cdot 2 \cdot \ldots \frac{q}{2}}.$$

To prove the result for cosines (not supplied by Babbage), if

$$z = \cos\phi + i\sin\phi,$$

$$\cos\phi = \tfrac{1}{2}[z + (1/z)],$$

$$\cos r\phi = \tfrac{1}{2}[z^r + (1/z^r)].$$

Then

$$\cos^{2n}\phi = (1/2^{2n})\left[z + \left(\frac{1}{z}\right)\right]^{2n}$$

$$= \frac{1}{2^{2n}}\left[z^{2n} + {}^{2n}C_1 z^{2n-2} + {}^{2n}C_2 z^{2n-4} + \cdots + {}^{2n}C_n + \cdots + \frac{1}{z^{2n}}\right]$$

$$= \frac{1}{2^{2n}}[2\cos 2n\phi + 2 \cdot {}^{2n}C_1 \cos(2n-2)\phi + \cdots + {}^{2n}C_n].$$

Using lemma 1 as in the previous case, this means that summing vertically, all the columns are zero except for exactly p terms, all equal to $1/2^q \cdot {}^qC_{q/2}$. Consequently the result follows since

$$p \cdot 1/2^q \cdot {}^qC_{q/2} = p/2^q \cdot \frac{q \cdot q - 1 \cdot \ldots q/2 + 1}{1 \cdot 2 \cdot \ldots q/2}.$$

Armed with these lemmas, it now becomes possible to prove many of Stewart's results with comparative ease. To take an example (39th proposition):

> Let there be any regular figure circumscribed about a circle, and let the number of sides be m, and let n be any number less than m; let r be the radius of the circle; and from any point in the circumference of the circle let there be drawn perpendiculars to the sides of the figure: the sum of the nth powers of the perpendiculars will be equal to
>
> $$m \times \frac{1 \cdot 3 \cdot 5 \ldots 2n-1}{1 \cdot 2 \cdot 3 \ldots n} r^n.$$

(*Ibid.*, p. 9.)

Babbage shows quite simply that the geometrical problem is equivalent to summing the series

$$2^n r^n \left[\left(\sin \frac{\theta}{2} \right)^{2n} + \left(\sin \overline{\frac{\theta}{2} + \frac{2\pi}{m}} \right)^{2n} + \cdots + \left(\sin \frac{\theta}{2} + \overline{m-1} \frac{2\pi}{m} \right)^{2n} \right],$$

and from the third lemma, the result follows at once.

In a similar way Babbage proves Stewart's more general propositions, numbers 40, 41 and 54.

Since the other two geometrical papers both concern the subject of porisms, it will be convenient to deal with the later one first as it attempts to define this concept, largely unfamiliar to modern mathematicians. The article is from the *Edinburgh Encyclopaedia*, 1830, **17**, 106–14. Babbage probably wrote the article much earlier than this, as correspondence from the editor, Sir David Brewster, indicates. The first of these letters dated 22 November, 1818, contains the invitation: 'I shall be very glad to receive from you the articles *Notation & Porisms*", and adds: 'It will not be wanted for a year at least.'[2] Brewster wrote again on 3 July, 1821: 'Our encyclopaedia will be completed in $2\frac{1}{2}$ years.'[3] What are porisms? According to Babbage in his article, 'Porisms are a species of proposition in geometry much employed by the ancients: they appear to have been highly valued, and to have assisted them greatly in their geometrical researches' (*ibid.*, p. 106).

It is hard to understand where Babbage obtained this information. Apart from the porisms of Euclid, to be considered at length in a moment, the only Greek mathematician said to have used porisms was Diophantus, and the references to this latter work are so brief as to give hardly any information about the nature of the concept. The only major reference to Euclid's lost book on porisms apart from a slight one by Proclus, is in the *Treasury of Analysis* of Pappus. According to Babbage (*Edinburgh Encyclopaedia*, **17**, p. 107), Pappus

> has mentioned the three books on porisms by Euclid, and has given thirty-eight geometrical propositions, of so very considerable difficulty, as useful for the comprehension of the work itself. These, together with the imperfect definition, and an example of a porism which refers to a figure that is lost, and which is so remarkably confused as almost to render its reconstruction impossible, are all the data that remained.

Babbage does not state this obscure example, but according to T. L. Heath it is

> If from two given points straight lines be drawn meeting on a straight line given in position, and one cut off from a straight line given in position (a segment measured) to a given point on it, the other will also cut off from another (straight line a segment) having to the first a given ratio.[4]

Heath's translation makes the proposition more understandable, but it throws no light on the central question of why this particular statement should be singled out to be called a porism. What is it that makes a porism distinct from any other geometrical theorem?

Pappus's work was first made known to the Western world with Commandine's translation in 1589. Mathematicians were fascinated by this reference to Euclid's lost work, and many attempts at restoration were made, including one by Girard, published in 1629. Fermat's posthumous publications contained a short paper: 'Porismatum Euclidoeorum renovata doctrina'. According to Babbage, Fermat produced some propositions which were definitely porisms, but failed either to restore Euclid's work or to give a satisfactory definition of a porism. A mathematician of Halley's calibre was baffled by Pappus's obscurity, for Babbage quotes him as admitting: 'Hactenus porismatum descriptio nec mihi intellecta nec lectori profutura.'[5]

The Scottish mathematician Robert Simson (1687–1768) made a further serious attempt in the *Philosophical Transactions*, 1723, to define a porism and restore Euclid's work. His definition certainly failed to resolve the obscurity and Playfair translated it from Simson's Latin as, 'A Porism is a Proposition in which it is proposed to demonstrate that one or more things are given, between which and every one of innumerable other things, not given, but assumed according to a given law, a certain relation, described in the Proposition, is to be shewn to take place.'[6] Playfair's own definition, quoted also in Trail's work, is quite distinct but much clearer: 'A Porism is a Proposition affirming the possibility of finding such conditions as will render a certain problem indeterminate, or capable of innumerable solutions.'[7] Trail actually disagrees with this definition on the grounds that it is too simple to have been lost from the whole of Greek mathematics.

Babbage, however, accepts Playfair's definition without question, and proceeds to give a series of his own geometrical results as illustrations.

To consider one of these, a circle is taken of radius r.

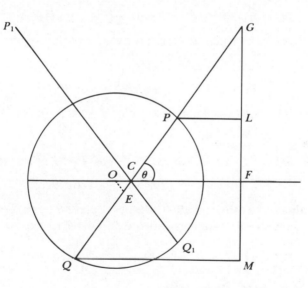

A point C is taken on a diameter such that $OC = v$, and a straight line FL drawn perpendicular to this diameter. It is required to find what angle a chord PQ passing through C should make with a given diameter, if the chord P_1Q_1 passing through C perpendicular to PQ, is such that the product of the perpendiculars from

P, Q to FL is equal to the product of the perpendiculars from P_1, Q_1 to FL.

Let the required angle PCF be θ, and OE be the perpendicular from O to QP. Then

$$CE = v \cos \theta,$$
$$OE = v \sin \theta$$

and therefore

$$CP = PE - CE = \sqrt{(r^2 - v^2 \sin^2 \theta)} - v \cos \theta$$

and

$$CQ = \sqrt{(r^2 - v^2 \sin^2 \theta)} + v \cos \theta.$$

Also

$$CG = (a - v)/\cos \theta, \text{ where } OF = a;$$

hence

$$PG = (a - v)/\cos \theta + v \cos \theta - \sqrt{(r^2 - v^2 \sin^2 \theta)},$$
$$QG = (a - v)/\cos \theta + v \cos \theta + \sqrt{(r^2 - v^2 \sin^2 \theta)},$$

and

$$PL = PG \cos \theta = a - v + v \cos^2 \theta - \cos \theta \sqrt{(r^2 - v^2 \sin^2 \theta)},$$
$$QM = QG \cos \theta = a - v + v \cos^2 \theta + \cos \theta \sqrt{(r^2 - v^2 \sin^2 \theta)},$$

and

$$PL \cdot QM = (a - v + v \cos^2 \theta)^2 - \cos^2 \theta (r^2 - v^2 \sin^2 \theta)$$
$$= (a - v)^2 + (2av - v^2 - r^2) \cos^2 \theta.$$

To evaluate the other product, simply replace θ by $(\pi/2) + \theta$, and since $\cos[(\pi/2) + \theta] = \sin \theta$, substitute $1 - \cos^2 \theta$ for $\cos^2 \theta$. Then

$$P_1L_1 \cdot Q_1M_1 = a^2 - r^2 - (2av - v^2 - r^2) \cos^2 \theta.$$

Putting $PL \cdot QM = P_1L_1 \cdot Q_1M_1$,

$$(a - v)^2 + (2av - v^2 - r^2) \cos^2 \theta = a^2 - r^2 - (2av - v^2 - r^2) \cos^2 \theta;$$

hence

$$2(2av - v^2 - r^2) \cos^2 \theta = 2av - v^2 - r^2,$$

from which $2\cos^2\theta = 1$ and $\cos\theta - \pm(1/\sqrt{2})$, giving the rather obvious solution $\theta = \pi/4$.

However, if a, v and r had been chosen so that $(a-v)^2 - a^2 + r^2 = 0$, so that $v = a \pm \sqrt{(a^2 - r^2)}$, then the equation to determine θ becomes indeterminate, and any value of θ satisfies it. This is the porism which Babbage enunciates finally as:

> A circle and a straight line being given in position, a point may be found within the circle, such, that if any two chords are drawn through that point at right angles to each other, the rectangle under the perpendiculars, let fall from the extremities of the first chord to the line given in position, shall always be equal to the rectangle under the perpendiculars let fall from the extremities of the other chord on the same line.[8]

Babbage ends this paper in the *Edinburgh Encyclopaedia* (p. 114) with an assertion reflecting the contemporary view of the usefulness of mathematics: 'It must be acknowledged that these truths, in their geometrical form, are useful only for the purpose of cultivating those mental habits which mathematical studies tend so strongly to promote.'

The subject of porisms was considered by other mathematicians of the nineteenth century, notably the Frenchman Chasles, but there has been no recent interest, owing to the declining importance of pure Euclidean geometry.

The third paper to be considered is 'On the application of analysis to the discovery of local theorems and porisms' from the *Transactions of the Royal Society of Edinburgh*, 1823, **9**, 337–52. It was read on 1 May 1820, but is signed 1 July 1818.

The paper really seems to be concerned with finding geometrical applications for the calculus of functions. The result is not very satisfying, and nearly all the theorems are worked backwards rather than proved. To supply one typical example (*ibid.*, p. 341), let ABC be any periodic curve for the second order $y = \alpha x$. Using Babbage's definitions from the calculus of functions, this means that α is such a function that $\alpha[\alpha(x)] = \alpha^2 x = x$. Then let DB, EC be any two corresponding ordinates. The problem is first to find another curve AFG such that the sum of its two ordinates at D and E is always constant. If $AD = x$, $AE = DB = \alpha x$ and, since $\alpha^2 x = x$, $EC = x$. If curve AFG is $y = \psi x$, $FD = \psi x$, $GE = \psi(AE) = \psi\alpha x$. The function ϕx must then satisfy the equation $\psi x + \psi\alpha x = c$, and Babbage quotes as solution $\psi x = c\phi x/(\phi x + \phi\alpha x)$ where ϕ is an arbitrary function.

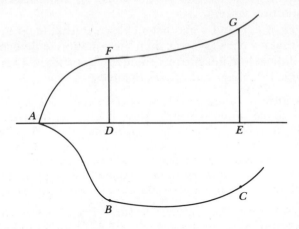

From this result, he deduces the not very interesting theorem that:

> Any of this family of curves being given, a periodic curve of the 2nd order (*ABC*) may always be found such, that if we take any two abscissae in the curve given, respectively equal to any two corresponding ordinates of the curve found, and draw ordinates to the curve, and if we prolong either of these ordinates (*EG*) above the curve, until the part above (*GH*) is equal to the first of the two ordinates, the extremity of the ordinate, thus increased, will always be situated in a right line given by position, Fig. 3. (*Ibid.*, p. 341.)

Fig. 3 is of course the diagram given above.

The other theorems in this paper are of a similar nature, and unlikely to interest anyone. It has to be concluded from the evidence of these three papers that Babbage was much more at home in the realm of analysis than in that of geometry.

NOTES

1. M. Stewart, *Some General Theorems of considerable use in the higher parts of Mathematics*, Edinburgh, 1746, p. i.
2. British Museum Additional Manuscripts 37182, No. 105.
3. *Ibid.*, No. 371.
4. T. L. Heath, *A Manual of Greek Mathematics*, Oxford, 1931, p. 264.
5. British Museum Additional Manuscripts 37182, No. 371, p. 107; quoting E. Halley, *Appolonias De Sectione Rationis*, Oxford, 1706, p. xxxvii.

6. W. Trail, *Account of the Life and Writings of Robert Simpson, M.D.*, London, 1822, p. 45.
7. *Ibid.*, p. 50.
8. British Museum Additional Manuscripts 37182, No. 371, p. 110.

7

Notation

It is clear that Babbage regarded the question of notation as one of supreme importance in mathematics. Good notation leads to rapid mathematical progress and poor symbolism to stagnation. The history of mathematics shows many examples of the correlation between advances made and the notation used. Babbage believed that in any type of logical reasoning it was essential to take great pains first to derive a notation both simple and comprehensive. Then not only was the immediate working of a problem greatly facilitated but even new and unsuspected results could be suggested. Sometimes a notation could be strong enough to open up new branches of a subject and greatly assist the whole process of mathematical discovery.

Babbage was, as we have seen, very much involved in notational reform in his early career, being born at a time when symbolism was at one of its lowest points in mathematical history. Later, he wrote three long and interesting papers on the subject of notation. These were 'Observations on the notation employed in the calculus of functions', *Transactions of the Cambridge Philosophical Society*, 1821; 'On the influence of signs in mathematical reasoning', *Transactions of the Cambridge Philosophical Society*, 1827, and an article 'On notation' for the *Edinburgh Encyclopaedia*, 1830.

The point about the relative merits of the differential and fluxionary notation which so exercised him in his undergraduate days is taken up again in the last of these (*Edinburgh Encyclopaedia*, 1830, **15**, 399):

> The superiority of the use of the d over the system of dots has become so very apparent, that the course of a few years will, in all probability, render the latter obsolete, a circumstance which induces us to forbear entering into any lengthened detail of

their comparative values, and to content ourselves with merely
indicating a few of the grounds for rejecting the latter: these
are, 1st, The uncertainty which may arise respecting the letters
to which the indices refer, and the confusion which arises from
having two indices to a letter. 2d, The want of analogy with
other established notations, such as those relating to the symbols Δ and δ. 3d, The great difficulty, if not the impossibility, of
representing, by their means, theorems relating to the separa-
tion of operations from quantities.

Of these three, the first reason is a matter of convenience, and
the other two are of a more mathematical nature. The inconvenience probably relates to the confusion of dots and dashes which
arise when different values of a function are differentiated any
number of times. There is no point in retaining a complicated
symbol when a simpler one will serve at least as well. Babbage's
second point refers to the operational nature of the symbol 'd', a
property which certainly would not have been noticed had 'dots'
remained. The third is a reference to the great difficulty which
would arise if the fluctionary notation had to be used in problems
involving differentiation, in the calculus of functions.

In the first book written by Babbage and his friends, called
Memoirs of the Analytical Society, Cambridge, 1813, the preface
contains some interesting remarks on notation in the context of a
historical survey of mathematics, as has already been discussed in
chapter 3.

Seven years later, Babbage wrote the first of his three papers
devoted to the subject of notation: 'Observations on the notation
employed in the calculus of functions', *Transactions of the Cambridge Philosophical Society*, 1821, **1**, 63–76. The paper was read on
1 May 1820, but dated 26 February 1820.

This is quite a remarkable paper, for the writer not only
demonstrates conclusively the excellence of his notation introduced for his own invented subject, the calculus of functions, but
even performs a series of calculations to prove the conciseness of
the symbolism. One might almost describe the paper as another
branch of mathematics invented by Babbage, the 'calculus of
notations'.

Before considering any of the calculations, I will quote Babbage's introduction at length:

> Amongst the various causes which combine in enabling us by
> the use of analytical reasoning to connect through a long

succession of intermediate steps the data of a question with its solution, no one exerts a more powerful influence than the brevity and compactness which is so peculiar to the language employed. The progress of improvement in leading us from the simpler up to the most complex relations has gradually produced new codes of shortening the ancient paths, and the symbols which have thus been invented in many instances from a partial view, or for very limited purposes, have themselves given rise to questions far beyond the expectations of their authors, and which have materially contributed to the progress of the science. Few indeed have been so fortunate as at once to perceive all the bearings and foresee all the consequences which result either necessarily, or analogically even from some of the simplest improvements.

The first analyst who employed the very natural abbreviation of a^2 instead of aa little contemplated the existence of fractional negative and imaginary exponents, at the moment when he adopted this apparently insignificant mode of abridging his labor, so great however is the connection that subsists between all branches of pure analysis, that we cannot employ a new symbol or make a new definition, without at once introducing a whole train of consequences, and in defiance of ourselves, the very sign we have created, and on which we have bestowed a meaning, itself almost prescribes the path our future investigations are to follow. (*Ibid.*, pp. 63–4.)

The symbolism introduced in the first place for convenience, opens up totally new possibilities for the branch of reasoning being considered. This is admirably illustrated by Babbage's example of the use of $a^2, a^3, a^4, \ldots$ for $aa, aaa, aaaa, \ldots$ suggesting the possibility of quantities with negative or fractional indices. One could cite the case of the differential notation in the calculus similarly suggesting possibilities that would be hidden forever in the fluctionary notation.

Babbage now applies this principle to the construction of a suitable notation for the calculus of functions.

He uses the letter f to denote the general functional operation, and suggests that a repetition of the operation involving f should be written as $f^2(x)$ rather than ffx. Similarly $f^3(x), \ldots f^n(x)$ will be written for $fff(x)$ and $fff^n \ldots f(x)$.

From this he produces the obvious, but important, identity (A) that $f^{n+m}(x) = f^n f^m(x)$, where n and m are whole numbers.

It is apparent, although Babbage does not make the point here, but implies it in his subsequent calculations, that the relation is a commutative one, and that $f^n f^m(x) = f^m f^n(x) = f^{n+m}(x)$.

The equation (A), which can be proved inductively whenever the indices are positive integers, can now be used to assign meanings to the function raised to a fractional, surd or negative index. All we have to do is to use the equation (A) as a definition as far as these quantities are concerned, and a whole range of new functions are introduced in a way logically continuous with those already known. 'The index n was now defined by means of the equation (A) and was said to indicate such a modification of the function to which it is attached that the equation shall be verified' (*ibid.*, p. 65).

Putting $n = 0$ in (A), it follows that $f^m(x) = f^0 f^m(x)$. Now put $y = f^m(x)$ and the equation becomes $y = f^0(y)$. Since y can be taken as any variable, this implies that the symbol f^0 is the unitary function, that is the function which leaves the variable unaltered. It is consequently independent of the particular operation f, and so, as Babbage points out, is analogous to the algebraic result $x^0 = 1$.

Another result can be obtained by putting $n = 1$, $m = -1$. Then $f^0(x) = f^1 f^{-1}(x)$ so $f^1 f^{-1}(x) = x$. '$f^{-1}(x)$ must therefore signify such a function of x, that if we perform upon it the operation denoted by f it shall be reduced to x' (*ibid.*, p. 66).

It could have been indicated here, and possibly Babbage omits it because it is so apparent, that by taking $n = -1$, $m = 1$, the result $f^{-1} f^1(x) = x$ could have been obtained so that f^{-1} commutes with f^1. However, he is more concerned at this point to resolve the ambiguity which might easily arise through the presence of the inverse function.

He quotes as an example of possible ambiguity the function $f(x) = x^n$. Let $r_1, \ldots r_n$ be the roots of $v^n - 1 = 0$, then $f^{-1}(x)$ may be any of the n functions $r_1 x^{1/n}, r_2 x^{1/n}, r_n x^{1/n}$, for if the operation f is performed on any of these, the result is x. Thus we have $f(f^{-1}x) = x$ for $f^{-1}(x)$ being any of these n functions.

On the other hand, $f^{-1}(fx) = r_i x$, where i can be any integer between 1 and n, depending on which function is taken for f^{-1}, and only one of these gives $f^{-1} fx = x$.

Consequently, we must define $f^{-1}(x)$ not as any function which has the property $f(f^{-1}x) = x$, but the one which at the same time makes $f^{-1}(fx) = x$. Only then can we call f^{-1} the inverse of f. 'It was more necessary to make this observation, because several errors have arisen from not attending to it, and because that particular form of f^{-1} which gives $f^{-1}fx = x$ possesses peculiar properties: $f^{-1}(x)$ is then the inverse function of fx; and if we have the equation $f(x) = y$, we may indicate its resolution thus $x = f^{-1}(y)$' (*ibid.*, p. 66).

Babbage does not consider the question of whether the inverse function is now unique or not. It can, however, easily be proved, assuming only that functions obey the associative law that if f, g, and h are any three functions, then $f[(gh)x] = fg[h(x)]$, an assumption which the writer unknowingly makes throughout his work on the calculus of functions.

Suppose then, that we have found an inverse function $f^{-1}(x)$ which has the properties $f[f^{-1}(x)] = x$ and $f^{-1}[f(x)] = x$; and also that there is a second inverse function $g(x)$ which has the properties $f[g(x)] = x$ and $g[f(x)] = x$. Consider the composite function $f^{-1}\{f[g(x)]\} = f^{-1}(x)$. By the associative law, $f^{-1}\{f[g(x)]\} = f^{-1}f[g(x)] = g(x)$; hence $f^{-1}(x) = g(x)$, and $f^{-1}(x)$ is unique.

Babbage now introduces the more complicated idea of the function of two variables, and develops the concept analogically to his work on the functions of a single variable.

Suppose we take a function of two variables, $\psi(x, y)$ and replace both the x and y by the function $\psi(x, y)$ itself, obtaining $\psi[\psi(x, y), \psi(x, y)]$. Babbage writes this as $\psi^{\overline{2.2}}(x, y)$, the bar indicating that the replacement is symmetrical with respect to x and y. He then makes the inductive assertion that if n and m are positive integers, then

$$\psi^{\overline{n,n}}[\psi^{\overline{m,m}}(x, y), \psi^{\overline{m,m}}(x, y)] = \psi^{\overline{n+m, \, n+m}}(x, y) \qquad (B)$$

analogous to result (A), and in a similar way takes this equation as a definition for other values of n and m. Putting $n = 0$, we have

$$\psi^{\overline{0,0}}[\psi^{\overline{m,m}}(x, y), \psi^{\overline{m,m}}(x, y)] = \psi^{\overline{m,m}}(x, y).$$

Then if $v = \psi^{\overline{m,m}}(x, y)$, $\psi^{\overline{0,0}}(v, v) = v$, defining the operator $\psi^{\overline{0,0}}(x, x)$ which again, turns out to be independent of ψ. Then putting $m = 0$, $n = 1$,

$$\psi^{\overline{1,1}}[\psi^{\overline{0,0}}(x, y), \psi^{\overline{0,0}}(x, y)] = \psi^{\overline{1,1}}(x, y).$$

Thus $\psi^{\overline{0,0}}(x, y)$ is such a function of x and y that when it is simultaneously substituted in $\psi(x, y)$ for x and y, it shall give the function $\psi(x, y)$.

It is easy to illustrate this when $\psi(x, y)$ is homogeneous in x and y; say $\psi(x, y) = x^2 + 2y^2$, then if $\psi^{0,0}(x, y) = k$, we have $\psi(k, k) = \psi(x, y)$, i.e. $k^2 + 2k^2 = x^2 + 2y^2$

$$k = \psi^{\overline{0,0}}(x, y) = \sqrt{[(x^2 + 2y^2)/3]},$$

and also

$$\psi^{\overline{0,0}}(v, v) = \sqrt{[(v^2 + 2v^2)/3]} = v,$$

confirming the previous result. This can obviously lead to ambiguity when different roots are considered. The problem is obviously very complicated, and Babbage is content to say of $\psi^{0,0}(x, y)$: 'It may have different values like all other inverse functions' (*ibid.*, p. 67).

The symmetrical substitution of functions for the variables within the function is the simplest type of such transformations when more than one variable is involved. The writer now turns his attention to non-symmetrical types.

If the function is substituted for x and the y value left unchanged, we have the notation $\psi^{2,1}(x, y)$ for $\psi[\psi(x, y), y]$, and proceeding by induction,

$$\psi^{n,1}[\psi^{m,1}(x, y), y] = \psi^{n+m,1}(x, y) \tag{C}$$

and again

$$\psi^{n,1}[x, \psi^{1,m}(x, y)] = \psi^{n,m+1}(x, y), \tag{D}$$

where n and m are positive integers. As usual we take (C) and (D) as definitions for other values of n and m. A miscellany of new functions quickly appear. Putting $n = 0$ in (C);

$$\psi^{0,1}[\psi^{m,1}(x, y), y] = \psi^{m,1}(x, y)$$

and replacing $\psi^{m,1}(x, y)$ by v, $\psi^{0,1}(v, y) = v$ or since v is a dummy suffix,

$$\psi^{0,1}(x, y) = x. \tag{E}$$

Similarly,

$$\psi^{1,0}(x, y) = y. \tag{F}$$

Now put $n = 1$, $m = -1$ in (D). Then

$$\psi^{1,1}[x, \psi^{1,-1}(x, y)] = \psi^{1,0}(x, y) = y.$$

This implies that if $\psi(x, y) = v$, then

$$y = \psi^{1,-1}(x, v) \quad \text{and} \quad x = \psi^{-1,1}(v, y). \tag{G}$$

This very satisfying result on inverse functions can be immediately generalised to functions of more than two variables, of which the following are typical results:

$$`\psi^{0,1,1,\cdots}(x, y, \ldots) = x \text{ etc.}$$

If $\psi^{1,1,1,\cdots}(x, y, z \ldots) = v$, $y = \psi^{1,-1,1,\cdots}(x, v, z \ldots)$' (*ibid.*, p. 68).

Having introduced this elegant notation, Babbage now proceeds to solve a series of problems on notation.

The first problem, on p. 68, is to determine how often the letter x would occur in the expression $\psi^{\overline{n,n}}(x, y)$ if written out in full.

He solves the problem by means of the calculus of differences. Suppose x occurs U_n times. Then in the equation

$$\psi^{\overline{n+1,n+1}}(x, y) = \psi^{\overline{n,n}}(\psi(x, y), \psi(x, y)),$$

x occurs U_{n+1} times on the left and $2U_n$ times on the right, since if $\psi^{\overline{n,n}}(x, y)$ were expanded, each x and each y would be replaced by an x when $\psi^{\overline{n,n}}[\psi(x, y), \psi(x, y)]$ is obtained from it. Hence $U_{n+1} = 2U_n$, which has solution $U_n = C \cdot 2^n$. Now if $n = 1$, $U_1 = 2C$ and $U_1 = 1$, then $C = \frac{1}{2}$, $U_n = \frac{1}{2} \cdot 2^n$, and x occurs 2^{n-1} times.

The second problem, on the same page, is to determine the number of times ψ occurs in the same expression expanded.

Let U_n be the number of times. Then each x in $\psi^{\overline{n,n}}(x, y)$ is replaced by a ψ, and so also is each y. From the previous problem, there are 2^{n-1} of each of these, making 2^n altogether.

We have the difference equation $U_{n+1} = U_n + 2^n$ and hence $U_n = 2^n + C$. If $n = 1$, $U_1 = 2 + C = 1$: $C = -1$; therefore ψ occurs $2^n - 1$ times in $\psi^{\overline{n,n}}(x, y)$.

The third problem (*ibid.*, p. 69) generalises the first, finding how often the symbol x occurs in the expansion of the function $\psi^{\overline{n,n,\cdots n}}(x_1, \ldots x_i)$.

Proceeding as in the first problem, the difference equation is $U_{n+1} = iU_n$, from which it can be shown that x occurs i^{n-1} times.

Problem IV on the same page is to find how often ψ occurs in the function of Problem III.

The difference equation obtained is $U_{n+1} = U_n + i^n$ which on integration gives the result that ψ occurs $(i^n - 1)/(i - 1)$ times.

The first four problems have all been on symmetrical transformations, and the remainder deal with the non-symmetrical cases.

Problems V and VI are very simple. Problem V (*ibid.*, p. 70) is to determine how many times x and y occur in the expression $\psi^{n,1}(x, y)$.

In changing from $\psi^{n,1}$ to $\psi^{n+1,1}$, x is replaced by $\psi(x, y)$. This means that the number of xs is unchanged by any such transformation. Therefore, there is only one x in $\psi^{n,1}(x, y)$. However, one extra y is introduced in the change from $\psi^{n,1}$ to $\psi^{n+1,1}$. This means that the appropriate difference equation is $U_{n+1} = U_n + 1$, from which $U_n = n + C$, if $n = 1$, $U_1 = 1 + C = 1$: $C = 0$ and y occurs n times.

In Problem VI on p. 70, it is found that ψ occurs n times in $\psi^{n,1}(x, y)$.

These results are now used to solve Problems VII and VIII, also, on p. 70, to evaluate the occurrence of x and y, and then ψ in the expression $\psi^{n,m}(x, y)$.

Problem VII is solved by considering the definition:

$$\psi^{n,m}(x, y) = \psi^{n,1}[x, \psi^{1,m-1}(x, y)].$$

From V, x occurs $m - 1$ times in $\psi^{1,m-1}(x, y)$. Replacing $\psi^{1,m-1}(x, y)$ by z, z occurs n times in $\psi^{n,1}(x, z)$ and x once, using V again. But x occurs $m - 1$ times in z, and consequently x will occur $(m - 1)n + 1 = mn - n + 1$ times in the whole expression.

y occurs once only in $\psi^{1,m-1}(x, y)$ and z occurs n times in $\psi^{n,1}(x, z)$; therefore y occurs n times in $\psi^{n,m}(x, y)$.

In Problem VIII it is shown that ψ occurs nm times in $\psi^{n,m}(x, y)$.

Problem IX (*ibid.*, p. 71) is a generalisation for more than two variables. The question is to find how often $x_1, x_2, \ldots x_k$ occur in $\psi^{a,b,c,\cdots}(x_1 \ldots x_k)$.

We first consider $\psi^{a,1,1,\cdots 1}(x_1 \ldots x_k)$ and let x occur U_a times. Since $\psi^{a+1,1,\cdots 1}(x_1, \ldots x_k) = \psi^{1,1,\cdots 1}[\psi^{a,1,\cdots 1}(x_1, \ldots x_k), x_2, \ldots x_k]$ from which it can easily be shown that x_1 occurs once and $x_2 \ldots x_k$ each a times in $\psi^{a,1,\cdots 1}(x_1, \ldots x_k)$.

Next $\psi^{a,b,\cdots 1}(x_1, \ldots x_k) = \psi^{a,1,\cdots 1}[x_1, \psi^{1,b-1,1,\cdots 1}(x_1, \ldots x_k), x_3, \ldots x_k]$ from which x_1 occurs $1 + a(b - 1)$ times, x_2 occurs a times, and $x_3, \ldots x_k$ occurs $a + a(b - 1) = ab$ times.

Babbage leaves the problem at this point, but it is obvious how it could be continued.

This completes the set of problems on the calculus of notations. In conclusion, he refers (*ibid.*, p. 76) to the economy supplied by this choice of notation, pointing out from the first two problems that the expression $\overline{\psi^{10,10}}(x, y)$, when expanded, contains x and y 512 times each, and ψ 1023 times, a total of 2047 letters comprehended in this very compact term. If a problem like, for example, finding functions ψ such that $\psi^{10,10}(x, y) = \psi(x, y)$ were to be solved, then 'If it were so developed it would require a much longer time merely to comprehend the enunciation of the problem than it would to understand and solve it in its contracted form.'

This paper provides a splendid example of Babbage's principles on notation. He has shown how a notation may be logically derived from the subject matter, and how its simplicity and conciseness can lead to a whole series of unsuspected results. In addition, we have this remarkable set of problems in which he has shown that the art of inventing notation can itself be reduced to a mathematical procedure. The great economy which this notation is shown to ensure is an admirable illustration of his contention that there should be as few signs of magnitude as possible.

The other two papers are more concerned with the principles governing the construction of a mathematical notation. The first of these, 'On the influence of signs in mathematical reasoning', was read on 16 December 1821, but not published until the second volume of the *Transactions of the Cambridge Philosophical Society* came out in 1827.

With his previous work on the subject, and a most interesting title that suggests that the writer is about to delve even more deeply into the nature of mathematics, the paper is most disappointing. It is much longer than the previous ones mentioned in this chapter, but the theme rambles over a whole range of mathematical topics, including both the calculus of functions and porisms, but without saying anything of great interest about any of them, and the mathematical level seems much lower than in the former paper, as if he were writing to mathematicians then, but to interested laymen now. It is noticeable throughout his writings that Babbage wrote papers at entirely different intellectual levels, but it is surprising that he could have written two such contrasting papers on the same subject for the same journal within eighteen months of each other. It is just possible that he was beginning to lose interest in pure mathematics out of preference for his great vision of the computer which began about this time. This is borne out by the fact that the four papers he published in 1822 all refer to the application of machinery to the calculation of mathematical tables, and this idea probably began to occupy his mind to the exclusion of earlier interests.

The paper begins with a reference to the history of the calculus. He points out that since its invention, research has tended to exploit the possibilities of the subject rather than to examine its principles, 'hence the lapse of nearly a century has been required to fix permanently the foundations on which the calculus of Newton and Leibnitz shall rest' (*ibid.*, pp. 325–6).

One has to bear in mind that this paper was written in 1821, the year that Cauchy's *Cours d'Analyse* was published, the book which, in conjunction with the same writer's later work, the *Résumé des Leçons sur le Calcul Infinitésimal*, Paris, 1823, and *Leçons sur le Calcul différentiel*, Paris, 1829, helped to establish the differential calculus on a well-defined logical basis, dependent on the concept of the limit. According to Babbage, the calculus was established within a century of its invention, which would place this, in his mind, well back into the eighteenth century and, in his opinion, by Lagrange, as we have seen in chapter 5.

Babbage goes on to speak of 'a profusion of notations which threaten, if not duly corrected, to multiply our difficulties instead of promoting our progress' (*Transactions of the Cambridge Philosophical Society*, **2**, 326). Few examples of these are given in this paper, but there are several in the following one. He proposes as a solution: 'As a remedy to the inconveniences which must inevitably result from the continued accumulation of new materials, as well as from the various dress in which the old may be exhibited, nothing appears so likely to succeed as a revision of the language in which all the results of science are expressed' (*ibid.*, p. 326).

This is always the case in mathematical sciences. The best way to ensure continuity of progress without confusion, when new advances are being made and old methods improved, is to strive for a notation which is simple, expressive and, above all, universal.

In continuing his discussion of notation, Babbage takes as an illustration the great advantages introduced into geometry by the use of analysis.

> The assumption of lines and figures to represent quantity and magnitude, was the method employed by the ancient geometers to present to the eye some picture by which the course of their reasonings might be traced: it was however, necessary to fill up

this outline by a tedious description, which in some instances even of no peculiar difficulty became nearly unintelligible, simply from its extreme length: the invention of algebra almost entirely removed this inconvenience, and presented to the eye a picture perfect in all its parts; disclosing at a glance, not merely the conclusion in which it terminated, but every step of its progress. (*Ibid.*, p. 330.)

It was probably this unhelpful notation which became as strong a reason as any other for the decline in Greek mathematics after the time of Archimedes and Apollonius. It was simply not possible for a mathematician following them to use such extraordinarily difficult methods and hope to discover results beyond those already established by these two great men. Similarly, the geometrical reasoning and notation of Newton proved too great a handicap for the mathematicians of lesser ability who tried to develop his work in Great Britain. Babbage points out rightly that the advent of analytical geometry has revolutionised the subject, making it possible to comprehend the whole argument at once and derive difficult results with great facility. He makes the further point that geometrical diagrams may lead to accidental errors, but algebraic symbols cannot. A third superiority of analytical over synthetic methods is well expressed in his next point: 'The signs used in Geometry, are frequently merely *individuals* of the *species* they represent; whilst those employed in Algebra having a connection purely arbitrary with the species for which they stand, do not force on the attention one individual in preference to any other' (*ibid.*, p. 338).

It is the nature of geometry to deal with particular cases one by one, and although the enumeration of these cases may suggest general results the subject lacks the power to establish them. Algebra, on the other hand, is designed to deal with general results of a problem and from these to work out easily any number of particular ones. The comparison between the simple results being laboriously compiled by geometrical methods, and then proved as special cases of a more algebraic result, is well illustrated in Babbage's own paper concerning Matthew Stewart's general theorems, as discussed in chapter 6.

On p. 346, he mentions the three stages in a problem where analysis is applied. They are:

I. The first stage consists in translating the proposed question into the language of analysis.

II. The second, comprehends the system of operations necessary to be performed, in order to resolve that analytical question
into which the first step had transformed the proposed one.
III. The third and last stage consists in retranslating the results
of the analytical process into ordinary language.

He remarks that the first of these is usually the stage at which
most errors occur. In the second it is important to bear in mind
the original nature of the problem so that the analysis does not
throw up some unexpected results. For example, in calculating
the length of a side of a regular heptagon inscribed in a circle, an
equation of the third degree is obtained leading always to three
real, distinct roots. There can, of course, only be one solution to
the problem and the strange ambiguity is resolved, if, lettering
the vertices $ABC\ldots$, it is realised that the three roots obtained
relate to the lengths AB, AC, and AD. The particular one
required can then be taken as the solution, but care must be taken
to see that in the process of analysis, the problem is not too greatly
generalised.

The next few pages in the paper refer to the subjects of porisms
and the calculus of functions, which I discussed in chapter 6; and
the next distinct mention of notation is on p. 361: 'Unlimited
variety in the use of signs is as much to be deprecated as too great
an adherence to one class of them; the one conceals the appear-
ance of relations that really exist, whilst the other, affecting to
display them too clearly, fails from its want of distinctness.' He
cites Lagrange as an example of the former type, as he used the
differential notation for calculus, invented the system of dashes
with functions, and then combined both of these with the dot
notation. To the modern mathematician this does not seem an
unreasonable use of notations.

Babbage goes on to encourage the use of suggestiveness and
symmetry in symbolism. For suggestiveness he commends the use
of v, v^1 for velocities, t, t^1 for times and s, s^1 for distances (*ibid.*, p.
368), the use of the symbols $a > b$, $b < a$ to denote inequalities
(*ibid.*, p. 370), and for symmetry the A, B, C, a, b, c notation for
the triangle invented by Lagrange (*ibid.*, p. 371). These principles, he points out, lead us back, curiously enough, to the origins
of language rather than being simply modern developments.

In Astronomy, this principle has been adopted with much
success, and signs $\odot$ and $\mathbb{C}$ for the Sun and Moon constantly
occur. It is rather singular that a principle on which the earliest

and most imperfect written language rested, should be found to add so essentially to the value of the most accurate and comprehensive: yet the language of hieroglyphics, is but the next stage in the progress of the art, to the mere picture of the event recorded; and the signs it employs, in most cases, closely resemble the things they express. In Algebra, although the principle has not been pushed to its extreme limits, the grounds of its observance are the same; the associations, are by its assistance, more easily and more permanently formed, and the memory most effectually assisted. (*Ibid.*, p. 374.)

This is, no doubt, helpful in an heuristic sense, but the intuitionism recommended in it must be deplored. An algebraic symbol means nothing except what it is defined to mean in terms of analysis, and to try to relate them to objects in the 'real' world can only lead to the sort of accidental errors which Babbage has pointed out may occur in geometry. Where, however, a simple, obvious and suggestive sign to represent a concept is available, it should certainly be used. In elementary algebra it is particularly helpful to use the signs referred to by the writer, but the more advanced parts of the subjects are, as Babbage would no doubt have anticipated, so general that it would be most unhelpful to introduce symbols with particular associations.

He concludes this paper:

I have now enumerated what appear to me to be the principle causes which exert an influence on the success of mathematical reasoning, and have illustrated, with examples, those which were susceptible of it. They may be recapitulated in a few words. The nature of the quantities which form the subject of the science, together with the distinctness of its definitions – the power of placing in a prominent light, the particular point on which reasoning turns – the quantity of meaning condensed into small space – the possibility of separating difficulties, and of combining innumerable cases, – together with the symmetry, which may be made to pervade the reasoning, both in the choice, and in the position of the symbols, are the grounds of that pre-eminence, which has been invariably allowed to the accuracy of conclusions deduced by mathematical reasoning. (*Ibid.*, p. 377.)

The conclusion is admirable, but the arguments and illustrations are to be found more in the other papers than in this one.

Babbage's final article, 'On notation', was an essay for volume 15 of David Brewster's *Edinburgh Encyclopaedia*, 1830. It is clear that the article was written a long time before this date. In a letter dated 3 July 1821, Brewster stated, 'Our encyclopaedia will be completed in $2\frac{1}{2}$ years . . .' and 'As we are very near *Notation* you would oblige me by sending it.'

A further letter from Brewster dated 25 February 1822, contains: 'I enclose a Proof of your Article on *Notation* in a Frank,'[1,2]

It seems unlikely that in view of his many other commitments, Babbage would have made many alterations to the essay, and it may be concluded that it was substantially completed by February 1822, so that the ideas are continuous with those of the previous paper.

The quality of the article is much superior to the previous one, and shows the writer in his best analytical form in clarifying the principles of notation.

He begins by defining notation as 'the art of adapting arbitrary symbols to the representation of quantities, and the operations to be performed on them' (*ibid.*, p. 394). He adds, on the same page, that 'the improvements thus introduced into notation, have in their turn directed the attention of the mathematician to wider and more general views of science.' This is followed by an historical development of notation which I have referred to in chapter 2. His purpose is to show that, originally, only brevity was the aim of introducing notation which then gradually developed into a more rationally planned language.

The major part of this article is, however, to supply general rules for those wishing to invent notations to express newly discovered relationships. For the sake of reference I have attached numbers to the rules he enunciates. They are as follows, taken from pp. 395–9.

1. *'That all notation should be as simple as the nature of the operations to be indicated will admit.'*

This, says Babbage, is too obvious for comment, and we will concur.

2. *'That we must adhere to one notation for one thing.'*

Exceptions to this rule are to be found in writers of the highest rank. Babbage again castigates Lagrange for using dashes in his *Théorie des Fonctions Analytiques*, Paris, 1796, and dots for explaining the calculus of variations in his *Leçons sur le Calcul des Fonctions*, Paris, 1806. This is rather petty. A mathematician today would use dots, dashes, small d's or large d's, whichever happened to be more convenient and could even, without confusion, combine these in the same problem. However, what is much less forgiveable is Lagrange's other notational error in the latter work, that of using the expression sin 1 continually, when he meant sin $(\pi/2)$. Another culprit was Legendre, who in his *Théorie des Nombres* used the ' = ' sign in two different senses, 'first, in its ordinary acceptation; and, secondly, he places it between two quantities, to denote, that when they are divided by the same quantity, they will have the same remainder'. This is the difference between equality and congruence modulo some integer. In view of this criticism, Babbage would doubtless be appalled to see the latter signified by the identity symbol today.

3. *'When it is required to express new relations that are analogous to others for which signs are already contrived, we should employ a notation as nearly allied to those signs as we conveniently can.'*

He comments that it was this sort of feeling which probably led Stifel into his enquiry concerning negative exponents. As an example, very familiar to Babbage, he suggests that since xx is denoted by x^2, so the successive functions $\psi\psi$ operating on a variable should be denoted by ψ^2. This is a proposal which has certainly not been adopted by mathematicians today, as we will see in a later example. There appears to be no good reason why it has not been accepted.

4. *'We ought not to multiply the number of mathematical symbols without necessity.'*

This again requires no comment.

5. *'Whenever we wish to denote the inverse of any operation, we must use the same characteristic with the index* -1.'

In view of modern usage, Babbage's comment is interesting. 'In the equation $x = \tan y$, the value of y, in terms of x is usually expressed $y = $ arc tan x; this is both long and inelegant. It should, in complying with the rule just laid down, be written thus: $y = \tan^{-1} x$.'

Both of these notations are used today. The one Babbage favours is subject to ambiguity in the way it is used, and the other, for this reason, is more in favour. He is right, however, that 'arc tan' is a long and inelegant expression, contravening the most elementary rule of notation. The ambiguity is, however, not a necessary one; it is introduced by our own lack of clarity in notation, a point which will be more forcibly made when considering Babbage's rule 11. For the inverse functions of tan, sin and cos, symbols such as nat, nis and soc might be considered more appropriate.

6. *'That every equation ought to be capable of indicating a law.'*

What Babbage means by this rather startling statement is that signs are not to be brought in unnecessarily but only when essential, and introduced in such a way as to effectually preclude future innovators from temptations to change them.

7. *'It is better to make any expression an apparent function of n, than to let it consist of operations n times repeated.'*

The example Babbage gives to illustrate this rule is the expression of the infinite series

$$\frac{1}{1^n}+\frac{1}{2^n}+\frac{1}{3^n}+\cdots$$

by means of a definite integral. This could be written as

$$\int\frac{\mathrm{d}x}{x}\int\frac{\mathrm{d}x}{x}\dots(n)\int\frac{\mathrm{d}x}{1-x}\binom{x=0}{x=1}.$$

Expressed in this way it is not certain whether n repetitions of the operation of integration are intended, or a product of n terms. The difficulty can be overcome by using a notation which turns the expression into a function of n, for example,

$$\left(\int\frac{1}{x}\right)^n\frac{x}{1-x}\,\mathrm{d}x^n \quad\text{or even}\quad \left(\mathrm{d}^{-1}\frac{1}{x}\right)^n\frac{x}{1-x}\,\mathrm{d}x^n.$$

> 'To any one acquainted with the integral calculus, this could not convey an incorrect idea, although presented for the first time, and without explanation.'

8. *'That all notation should be so contrived as to have its parts capable of being employed separately.'*

This is another fairly obvious principle which the writer, with his great knowledge of contemporary mathematical texts, admits to be so universally adhered to, that it is not possible for him to find any examples of its neglect.

He now makes a very interesting general remark which probably has wider application than to the principles of mathematical notation.

> How frequently does it happen, even to the best informed, that they prefer one thing and reject another, from some latent

sense of their propriety or impropriety, without being immediately able to state the reasons on which such choice is founded; yet it cannot be doubted, when the selection appears to be the result of good taste, that it is guided by unwritten rules, themselves the valued offspring of long experience.

One always feels that there are inexpressible rules which govern choice and it is not at all easy to discover what they are. It is quite possible to have an aesthetic feeling for mathematics and prefer one thing and reject another out of a taste based on experience. These feelings do, however, depend on explicit rules as Babbage says, and it is a most praiseworthy achievement on his part, to be able to analyse a flair for good notation into definite working rules. If someone like Babbage could apply a similar talent to the strictly non-mathematical world, life would be much easier for the mathematically minded man.

9.   *'All letters that denote quantity should be printed in Italics, but all those which indicate operations, should be printed in a Roman character.'*

Babbage admits the difficulty which arises out of the use of Greek letters in mathematics. This principle has been universally ignored, but it would be useful to have some means of distinguishing between different categories of expression. One example of this is in the type of computer programming, invented by Babbage himself in which operations are denoted by a letter and locations by a number.

He makes at this point a useful distinction between functional characteristics and derivative characteristics. A functional characteristic is something like the expression $f(1 + x^2)$, where the function f is quite independent of the $1 + x^2$. Whatever filled the bracket, the nature of the function would be unchanged. However, in a derivative characteristic like $d(1 + x^2)$, the effect of the operation d depends entirely on the form of the function enclosed. This distinction leads to the next rule:

10.   *'Every functional characteristic affects all symbols which follow it, just as if they constituted one letter.'*

He goes on to say that in the operations of calculus, second and higher derivatives $(dx)^2$, $(dx)^3$, . . . $(dx)^n$ frequently occur, whilst we seldom meet with things like $d(x^2)$, $d(x^3)$ . . . Consequently, universal usage sanctions the use of dx^2, dx^3, . . . for the first type, while if the others are required, then they should be denoted by $d \cdot x^2$, $d \cdot x^3$, ...

The last rule is interesting. It is:

11. *'Parentheses may be omitted, if it can be done without introducing ambiguity.'*

The above work is an example of this.

As a further illustration, Babbage produces five ways which are used to represent the same quantity. These are:

'$(\text{sin. } \theta)^2$ $\text{sin. } \theta^2$ $\sin^2 \theta$ $(\sin \theta)^2$ $\sin \theta^2$'.

His comment will come as a surprise to modern mathematicians.

> 'The third is by far the most objectionable of any, and is completely at variance with strong analogies. An index in the situation in which it there occurs, invariably denote repetition of the symbol to which it is attached: in this case it would therefore mean $\sin \cdot \sin \theta$, or the sine of the sine of the arc θ.'

Babbage is quite correct. If two successive operations of the same function are to be written not as ffx but as f^2x, it follows that $\log(\log x)$ should be $\log^2 x$ and $\sin(\sin x)$ should be $\sin^2 x$. His notation seems a very logical one, but because it has not been accepted we use the unsatisfactory $\sin^2 x$ to denote $(\sin x)^2$ or $\sin^n x$ for $(\sin x)^n$. Our use creates an unnecessary ambiguity for this notation and means that $\tan^{-1} x$ is the same as $1/\tan x$ or $\cot x$ when we want it also to mean the inverse function of $\tan x$. As a result, we have to drop the expression $\tan^{-1} x$ in its more natural sense and revert to the clumsy arc tan x. I believe that a consistent use of Babbage's rules 3, 5 and 11 would improve the notation used today, and eliminate the ambiguities which are quite unnecessary.

To those who would say that $\sin^2 \theta$ means $(\sin \theta)^2$ because it is defined to be so, Babbage replies, 'Although a definition cannot be false, it may be improper; and the impropriety may arise either from its inducing, ambiguity, or from its offending against received principles; both which objections occur in the present instance.'

We see that Babbage was so steeped in the ideas of notation that he was able to adapt the principles in a practical way when it came to the construction of his engines.

> In the description of machinery by means of drawings, it is only possible to represent an engine in one particular state of its action. If indeed it is very simple in its operations, a succession

of drawings may be made of it in each state of its progress which will represent its whole course; but this rarely happens, and is attended with the inconvenience and expense of numerous drawings. The difficulty of retaining in the mind all the contemporaneous and successive movements of a complicated machine, and the still greater difficulty of properly timing movements which had already been provided for, induced me to seek for some method by which I might at a glance of the eye select any particular part, and find at any given time its state of motion or rest, its relation to the motions of any other part of the machine, and if necessary trace back the sources of its movement through all its successive stages to the original moving power. I soon felt that the forms of ordinary language were far too diffuse to admit of any expectation of removing the difficulty, and being convinced from experience of the vast power which analysis derives from the great condensation of meaning in the language it employs, I was not long in deciding that the most favourable path to pursue was to have recourse to the language of signs. It then became necessary to contrive a notation which ought if possible to be at once simple and expressive, easily understood at the commencement, and capable of being readily retained in the memory from the proper adaptation of the signs to the circumstances they are intended to represent.[3]

NOTES

1. British Museum Additional Manuscripts 37182, No. 371.
2. *Ibid.*, No. 407.
3. *Philosophical Transactions*, 1826, **116**, Part II, 250–1.

8

Babbage and his computers

There is no doubt that Babbage's major life interest was in the invention and construction of his calculating engines. He began work on these machines in 1819 and continued actively with few breaks until his death in 1871. These calculators, particularly, as will be shown, the analytical engine, embody all the important components of a modern computer; there was no field in which Babbage made such a vitally important contribution to science, and in which his ideas were so far ahead of his contemporaries. On the other hand, the engines meant to their creator immense toil, frustration, financial loss, and in the end failure to complete any of the ambitious projects he had intended.

Babbage invented his engines at a time when the concept of mechanical calculators had not gone beyond the idea of adding machines, with contrivances that could multiply by means of continual addition and were usually worked by turning a handle. It was also a time in which precision in mechanical engineering was certainly not of a high enough order to make his own very elaborate engines a viable proposition. Consequently, Babbage was faced at all times with the double problem, mathematical and mechanical, of not only originating the whole concept of automatic computing but also constructing a suitable machinery that would be precise enough to put this into practice. As a by-product of his ideas on computation, Babbage succeeded in setting a new standard of precision for the mechanical engineering industry, inventing a mechanical notation for the working part of his engines, and making many valuable observations and suggestions concerning mass production.

He also lived at a time when scientific research was almost entirely an amateur pursuit, and, in order to finance such an

173

ambitious project as the first 'difference engine', had a lively if ultimately fruitless battle with successive Governments of this country between 1823 and 1848. Babbage finally decided from his experiences that he would do better to proceed on an amateur basis, but his struggle set the precedent for future relations between government finance and scientific research.

Babbage had two major conceptions of an automatic calculator, the difference engine and the analytical engine, the latter being much more advanced than the former, and for the purposes of this chapter I will try to deal with each separately, and then attempt an assessment of their capabilities in relation to modern computers.

THE DIFFERENCE ENGINE

Babbage's earliest recollections of his original ideas for an automatic computer are contained in his autobiography, *Passages from the Life of a Philosopher*, London, 1864, written, it must be admitted, fifty years after the event, so that there is the possibility of inaccuracies. In an anecdote recalled to his memory by his friend, the Rev. Dr Robinson, the Master of the Temple, and one of the original members of the Analytical Society, he states:

> One evening I was sitting in the rooms of the Analytical Society, at Cambridge, my head leaning forward on the Table in a kind of dreamy mood, with a Table of logarithms lying open before me. Another member, coming into the room, and seeing me half asleep, called out, 'Well, Babbage, what are you dreaming about?' to which I replied 'I am thinking that all these Tables (pointing to the logarithms) might be calculated by machinery.'[1]

He adds that this incident must have happened in 1812 or 1813. Babbage certainly did not develop any of these ideas consciously for the next few years, when his scientific interests appear to have been restricted to pure mathematics. He returned to the theme of mechanical calculation again in 1819.

> About 1819 I was occupied with devising means for accurately dividing astronomical instruments, and had arrived at a plan which I thought was likely to succeed perfectly. I had also at that time been speculating about making machinery to compute arithmetical tables.

One morning I called upon the late Dr. Wollaston, to consult him about my plan for dividing instruments. On talking over the matter, it turned out that my system was exactly that which had been described by the Duke de Chaulnes, in the Memoirs of the French Academy of Sciences, about fifty and sixty years before. I then mentioned my other idea of computing Tables by machinery, which Dr. Wollaston thought a more promising subject.

I considered that a machine to execute the more isolated operations of arithmetic, would be comparatively of little value, unless it were very easily set to do its work, and unless it executed not only accurately, but with great rapidity, whatever it was required to do.

On the other hand, the method of differences supplied a general principle by which *all* Tables might be computed through limited intervals, by one uniform process. Again, the method of differences required the use of mechanism for Addition only. In order, however, to insure accuracy in the printed Tables, it was necessary that the machine which computed Tables should also set them up in type, or else supply a mould in which stereotype plates of these Tables could be cast.

I now began to sketch out arrangements for accomplishing the several partial processes which were required.[2]

In this way Babbage conceived the idea and then constructed his first difference engine which not only could perform calculations automatically but could even print out the results. This was a major breakthrough in the whole idea of mathematical machinery. The mathematical principle, which will be explained shortly, was a relatively simple one, although never previously used in a mechanical calculator. Babbage started to construct a small machine on these lines in 1820 and completed it in June 1822.

He first announced the successful completion in a paper read to the Astronomical Society of London on 14 June 1822, which was entitled 'Note on the Application of Machinery to the Computation of Astronomical and Mathematical Tables' He begins:

It is known to several of the members of this society that I have been engaged during the last few months in the contrivance of machinery, which by the application of a moving force may calculate any tables that may be required. I am now able to acquaint the society with the successful results at which I have arrived; and although it might at the first view appear a bold undertaking to attempt the construction of an engine which should execute operations so various as those which contribute

Plate 1. Ada Augusta King, Countess of Lovelace

to the formation of the numerous tables that are constantly required for astronomical purposes, yet to those who are acquainted with the method of differences the difficulty will be in a considerable degree removed.[3]

He adds that he has been able to use the machine in order to compute tables of squares, triangular numbers and successive values of the polynomial $x^2 + x + 41$, a quadratic function which particularly fascinated Babbage owing to its first thirty-nine values all being prime numbers.

It appears that he was not at this stage able to supply the machine with the means for printing out its own results for he mentions that the results of the latter quadratic are produced 'almost as rapidly as an assistant can write them down'. Also he adds that 'I have contrived methods by which type shall be set up by the machine in the order determined by the calculation.'[3]

Babbage next wrote a more detailed account of his work on this difference engine in a letter entitled 'On the Application of Machinery to the Purpose of Calculating and Printing Mathematical Tables', dated 3 July 1822, to Sir Humphry Davy, President of the Royal Society.

This letter is reproduced in a collection of material, relevant to the history of Babbage's calculating engines, compiled by his son, Major General H. P. Babbage – *Babbage's Calculating Engines*, London, 1889.

There are four points of particular interest in this letter. The first is an enlargement on the previous commentary on the computation of the successive values of $x^2 + x + 41$ in which the time factor is introduced. Babbage was concerned throughout his fifty years' association with computers that the time for basic operations should be successively reduced. Indeed as he pointed out, there is no object in constructing such a machine unless it can perform calculations very much more rapidly than a human calculator. The printing device not yet having been added, all the results had to be written down by an observer. Babbage remarks that at first he was able to keep pace with the engine, but became harder pressed when four-figure numbers were produced. He continues,

> In another trial it was found that thirty numbers of the same table were calculated in two minutes and thirty seconds: as these contained eighty-two figures, the engine produced thirty-three every minute.

In another trial it produced figures at the rate of forty-four in a minute. As the machine may be made to move uniformly by a weight, this rate might be maintained for any length of time, and I believe few writers would be found to copy with equal speed for many hours together.[4]

This is, of course, very slow compared with electronic standards, but an excellent start using mechanical means only. Unfortunately, Babbage never supplied a time study of any other machine, so that its performance might be compared, but he makes the point that by increasing the number of differences (the constructed machine dealing only with second order differences) the advantage in speed over the human operator becomes much more pronounced. I might add that working arithmetically I managed to compute the required eighty-two figures of the first thirty terms of $x^2 + x + 41$ in two minutes ten seconds, but would doubt my chances against a comparable fourth difference engine.

The second point of interest in this letter is a description of some of the tabulation possibilities of an enlarged difference engine. Babbage gives an analysis, which he repeats elsewhere in his writings and which obviously had great influence on him in designing the engine, of some calculations to produce new tables by order of the French Government. The formidable task entailed the computation of such tables as the logarithms of numbers from 1 to 10 000, calculated to nineteen decimals; the natural sines of each 10 000th of the quadrant, calculated to twenty-five figures; and the logarithms of the ratios of the sines to their arcs of the first 5000 of the 100 000ths of the quadrant, calculated to fourteen decimals. One table alone, the logarithms of numbers, was to contain about eight million figures. The tables were all computed by the method of differences, and Babbage was particularly interested in the way in which the labour was divided. According to him,

The calculators were divided into three sections. The first section comprised five or six mathematicians of the highest merit, amongst whom were M. Prony and M. Legendre. These were occupied entirely with the analytical part of the work; they investigated and determined the formulae to be employed.

The second section consisted of seven or eight skilful calculators habituated both to analytical and arithmetical computations. These received the formulae from the first section, converted them into numbers, and furnished to the third section the proper differences at stated intervals.

They also received from that section the calculated results, and compared the two sets, which were computed independently for the purpose of verification.

The third section, on whom the most laborious part of the operations devolved, consisted of from sixty to eighty persons, few of them possessing a knowledge of more than the first rules of arithmetic: these received from the second class certain numbers and differences, with which, by additions and subtractions in a prescribed order, they completed the whole of the tables above mentioned.[5]

Babbage makes the point that the ninety-six calculators could be reduced, largely by the elimination of the third section, to at most a dozen, if a large enough difference engine were introduced. He clearly saw the possibilities of such a machine for reducing routine human labour and saving cost. There are some commentators who make the point that by his analytical investigation of the pin-making industry and the penny-post system, Babbage was the inventor of the modern technique of 'operations research'. If one carries this labelling process to its logical conclusion, it might be claimed that Babbage in his study of the replacement of human labour by a computer in the French tables calculation, invented the still more recent discipline of 'systems analysis'!

Thirdly, it is interesting to see that even at this early stage after only two years with a difference engine, Babbage was already thinking about superior types of computers. He had in mind the prototype for a machine that multiplies directly without a continual addition: 'Of a machine for multiplying any number of figures (m) by any other number (n), I have several sketches; but it is not yet brought to that degree of perfection which I should wish to give it before it is to be executed.'[6]

The principle of differences normally requires addition as its only mathematical operation. To introduce an automatic multiplier would considerably extend the type of difference equation that could be solved.

He continues: 'Another machine, whose plans are much more advanced than several of those just named, is one for constructing tables which have no order of differences constant.'[6] Again, it is conceivable that Babbage was referring to a more advanced difference engine, but there is just the possibility, which we will investigate when considering H. P. Babbage's account of his father's thought, that the analytical engine was already in his mind.

Fourthly, there is the significant ending to the letter. 'Whether I shall construct a larger engine of this kind, and bring to perfection the others I have described, will in a great measure depend on the nature of the encouragement I may receive.'[7] This proved to be a singularly prophetic utterance in view of all the trouble which Babbage was destined to face a few years later in his attempts to obtain adequate financial support for his work.

His last paragraph is even more significant:

> Induced by a conviction of the great utility of such engines, to withdraw for some time my attention from a subject on which it has been engaged during several years, and which possesses charms of a higher order, I have now arrived at a point where success is no longer doubtful. It must, however, be attained at a very considerable expense, which would not probably be replaced by the works it might produce, for a long period of time, and which is an undertaking I should feel unwilling to commence, as altogether foreign to my habits and pursuits.[7]

This is tangible evidence of Babbage's personal unwillingness to be involved in the huge task of completing these more advanced engines. The subject with 'charms of a higher order' may well have been the calculus of functions, which as I have explained in chapter 4 was almost completely originated by Babbage in 1815 and abandoned in 1820 without an attempt to solve some of the major problems set by himself. In his *Passages from the Life of a Philosopher*, he remarks of the calculus of functions, 'This was my earliest step, and is still one to which I would willingly recur if other demands on my time permitted.'[8]

Brilliant as his engineering turned out to be, with the by-products already mentioned, it seems extraordinary that a man of Babbage's genius should have had to be condemned to spend so many years supervising the mechanical construction of his computers. This was another part of the price he had to pay for his conception of such calculators. He was so convinced of the potential of his engines, coupled with the realisation that no other person could do justice to the supervision of their construction, that he entered involuntarily into the years of frustration ahead.

The final paragraph of this letter to Davy also served as an answer to those who slanderously accused him of personal gain out of his subsequent development of the difference engine.

It would be appropriate at this stage to discuss the mathematical principle of the difference engine. The particular one that had been constructed, depended on the simple rule

$$\Delta^2 u_n = c, \text{ where } c \text{ is a constant.}$$

This simple difference equation has the general solution $u_n = \frac{1}{2}cn^2 + an + b$, where a and b are arbitrary constants.

If now u_0 and u_1 or u_0 and Δu_0 are given, then a and b are determined uniquely. This means that given the three constants c, u_0 and Δu_0, the successive values of any quadratic function can be calculated. To take $x^2 + x + 41$ as an example, the three given constants are $c = 2$, $u_0 = 41$ and $\Delta u_0 = 0$.

The sequence of values is then calculated quite simply from a table of differences as

n	u_n	Δu_n	$\Delta^2 u_n$	
0	41	0	2	
1	43	2	2	
2	47	4	2	
3	53	6	2	
4	61	8	2	etc.

The difference engine then consists of three columns of numbers starting at the values 41, 0, 2. The sequence of operations is that the third column is added to the second column, the second to the first and the result recorded; the third added to the second, the second to the first, the result recorded and the whole process repeated as many times as are required.

The great advantages of this simple principle for mechanical calculation are that the only operation required is addition, successive values for a table may be computed with great rapidity, and the solution may be checked at any stage, with the virtual guarantee that if this result were correct so also were all the previous ones.

As far as the construction of tables was concerned, all functions like log, sin, tan etc. could be very closely approximated to by a suitable polynomial expansion. If the degree of this polynomial were m, then a difference equation $\Delta^m u_n = c$ with the required constants has this particular polynomial as solution. Consequently a difference engine with $m + 1$ columns could compute the table.

Before embarking on the construction of such an engine with a high order of differences, Babbage found a certain fascination in new types of sequences that could be computed by the existing machine. He was interested enough to communicate two papers on his findings: the first (from a letter to Brewster dated 6 November 1822) was published as 'On the theoretical principles of the machinery for calculating tables' in Brewster's *Edinburgh Journal of Science*, 1822, and the second as 'Observations on the application of machinery to the computation of mathematical tables' in the *Memoirs of the Astronomical Society*, 1822, **1**, 311–14, dated 13 December 1822. Babbage made the same point in each of these, that it was quite easy to set up his engine to compute sequences of the form

$$\Delta u_z = \text{units fig. of } u_z$$

and

$$\Delta^2 u_z = \text{units fig. of } u_{z+1}$$

the latter resulting in the successive values 2, 2, 4, 10, 16, 28, 48, ... Babbage observes that he had never previously seen a treatment of such sequences and adds that it is remarkable that a piece of machinery can actually invent new mathematics.

These sequences have no intrinsic mathematical significance and serve as an illustration of one of Babbage's weaknesses that he could be easily sidetracked from a major task into a consideration of relatively unimportant matters.

This first engine was entirely successful in its performance and in bringing acclaim to its maker. The Astronomical Society of London presented a gold medal to Babbage on 13 July 1823, and this may have given him the encouragement he needed to go ahead with the task of constructing a much enlarged difference engine.

At this point, it is useful to look at the history of the financial problems associated with this construction. The facts are related in *Passages from the Life of a Philosopher*, the sixth chapter consisting of a statement relating to the difference engine drawn up by Sir Harris Nicolson, of whom Babbage says: 'I believe every paper I possessed at all bearing on the subject was in his hands for several months.'[9] There is also a detailed account in C. R. Weld's *History of the Royal Society*, London, 1848, vol. II, chapter XI, and all the relevant letters are to be found in the unpublished correspondence in the British Museum.

The two accounts agree that a copy of Babbage's letter to Sir Humphry Davy was sent to the Treasury and that the Lords of the Treasury requested an opinion from the Royal Society on 1 April 1823. The Royal Society on 1 May 1823 reported to the Treasury that

> Mr. Babbage has displayed great talent and ingenuity in the construction of his Machine for Computation, which the Committee thinks fully adequate to the attainment of the objects proposed by the inventory; and they consider Mr. Babbage as highly deserving of public encouragement, in the prosecution of his arduous undertaking.[10]

He had also sent a copy of this letter to his friend Olinthus Gregory who replied on 16 July 1822:

> The application of machinery to the purposes of computation, in the way you have so happily struck out is highly interesting, and cannot fail, I should think, to be exceedingly beneficial. I trust that our valued friend Mr. D. Gilbert, and some other friends to science who possess influence in high quarters, will exert it cordially on this occasion, and obtain an adequate grant from Government to complete and render extensively effectual the whole of your curious inventions.[11]

In July 1823, Babbage had an interview with Frederick Robinson, the Chancellor of the Exchequer, in which it was agreed that the difference engine, in view of its importance for compiling tables, was a project worthy of Government support. Unfortunately, no minute of this conversation was made, and both parties were afterwards uncertain of the exact nature of the agreement made. Babbage was under the impression that through receiving financial support, the engine became the property of the Government. The only tangible outcome was that Babbage was given £1500 from the Civil Contingencies, and he believed at the time that he would be able to provide the additional £3000–5000 required out of his private funds to complete the project in, at most, three years. The most unsatisfactory thing was that soon after this interview, a letter was sent from the Treasury to the Royal Society informing them that the Lords of the Treasury:

'Had directed the issue of £1,500 to Mr. Babbage, to enable him to bring his invention to perfection, in the manner recommended.'[12] The last four words 'in the manner recommended' caused a great deal of misunderstanding owing to the vagueness of the statement.

In July 1823, work began on the construction of the difference engine, and continued steadily for four years. To give some idea of the early progress, an article 'Mr. Babbage's calculating machine' written by F. Baily, appeared in the *Edinburgh Journal of Science*, 1824.

Baily reported that the machine under construction was intended to contain apparatus for printing results and also the device for direct multiplication mentioned in the letter to Davy. Concerning the mathematical principle, he states, 'The machine which he is calculating will tabulate the equation $\Delta^4 u_z = c$; consequently there must be a means of representing the given constant C, and also the four arbitrary ones introduced in the integration. There are five axes in the machine, in each of which one may be placed.'[13]

He adds that such a machine could solve difference equations of the type $\Delta^5 u_z = a u_{z+1}$, $\Delta^5 u_{z+1} = a u_z + \Delta^4 u_z$ etc. and that a table of logarithms could normally be computed by the equation $\Delta^4 u_z = c$ and a table of sines by $\Delta^2 u_z = a u_{z+1}$.

It appears that at this time Baily's ideas of the extent of the machine were limited. Babbage states in his autobiography that 'If the whole Engine had been completed it would have had six orders of differences, each of twenty places of figures, whilst the three first columns would each have had half a dozen additional figures.'[14] He later underlines this by referring to the mathematical principle of the machine as simply $\Delta^7 u_z = 0$ indicating the six orders of differences.

In any case, bearing in mind Baily's remarks about the equations for computing logarithms and sines, it is likely that if Babbage's difference engine had been completed it would have been able to reproduce all the work of the French tabulators without being fully extended.

By 1827, the expenses incurred in the construction were far greater than anticipated. In this year Babbage had a series of personal tragedies in which his father, his wife and two of his children all died, his own health broke down and he was medically advised to travel on the Continent. On his return to England towards the end of 1828, he approached the Government, headed by the Duke of Wellington, for further assistance. The Duke requested the Royal Society to enquire as to whether the progress of the machine was sufficient to justify their earlier confidence in its great possibilities.

The Royal Society set up a committee consisting of: 'The President and Secretaries, Mr. Herschel, Mr. Warburton, Mr. F. Baily, Mr. Barton, Comptroller of the Mint, Mr. Brunel, F.R.S., civil engineer, Mr. Donkin, civil engineer, Mr. Penn, and Mr. G. Rennie, F.R.S., civil engineer.'[15] In the minutes of the Council of the Royal Society, dated 12 February 1829, it was reported:

> From the result of whose examination of the drawings, the tools employed, and the work already executed, as detailed in the annexed Report, they have not the slightest hesitation in pronouncing their decided opinion in the affirmative.
>
> The Council of the Royal Society cannot conclude without stating their full concurrence in the Report of their Committee, comprising, as it does, among its members several of the first practical engineers and mechanicians in the country; nor without expressing a hope that while Mr. Babbage's mind is intently occupied on an undertaking likely to do so much honour to his country, he may be relieved, as much as possible, from all other sources of anxiety.[15]

In the annexed Report it is stated that they examined over 400 square feet of very fine drawings, and Babbage estimated that the drawing stage was nine-tenths completed. They were particularly impressed with the 'mechanical notation' invented by Babbage, which enabled the complementary parts of the machine to be compatible in practice. The committee suggested that the work would take a further three years to complete.

As a result of this Report, a Treasury Minute dated 28 April 1829 directed a further payment of £1500 to Babbage. This was hopelessly inadequate. According to Nicolson, the amount already expended on the engine was nearly £6700 of which £3000 had been supplied by the Treasury, the remainder by Babbage himself. The unequivocal report of the Royal Society, together with their desire that he should be relieved of financial anxiety, had been largely unheeded.

On the advice of his brother-in-law, Wolryche Whitmore, a meeting of Babbage's friends, consisting of Whitmore himself, The Duke of Somerset, Lord Ashley, Sir John Franklin, Dr

Fitton, Mr Francis Baily and Mr John Herschel, took place on 12 May 1829. They resolved that:

1st. That Mr. Babbage was originally induced to take up the work, on its present extensive scale, by an understanding on his part that it was the wish of the Government that he should do so, and by an advance of 1,500 l., at the outset; with a full impression on his mind, that such further advances would be made as the work might require.

2nd. That Mr. Babbage's expenditure had amounted to nearly 7,000 l., while the whole sum advanced by the Government was 3,000 l.

3rd. That Mr. Babbage had devoted the most assiduous and anxious attention to the progress of the Engine, to the injury of his health, and the neglect and refusal of other profitable occupations.

4th. That a very large expense remained to be incurred; and that his private fortune was not such as would justify his completing the Engine, without further and effectual assistance from Government.

5th. That a personal application upon the subject should be made to the Duke of Wellington.

6th. That if such application should be unsuccessful in procuring effectual and adequate assistance, they must regard Mr. Babbage (considering the great pecuniary and personal sacrifices he will then have made; the entire expenditure of all he had received from the public on the subject of its destination; and the moral certainty of completing it, to which it was, by his exertions, reduced) as no longer called on to proceed with an undertaking which might destroy his health, and injure, if not ruin, his fortune.

7th. That Mr. Wolryche Whitmore and Mr. Herschel should request an interview with the Duke of Wellington, to state to his Grace these opinions on the subject.[16]

The Duke of Wellington, instead of referring to the Report of the Royal Society, expressed the wish to inspect Babbage's work himself. In company with Goulburn, the Chancellor of the Exchequer and Lord Ashley, he visited Babbage in November, 1829, to see the model of the engine, the drawings and the parts in progress. This visit resulted in a further government offer of £3000, and the start of a close friendship between Babbage and the Duke.

Before accepting this money, Babbage wrote to Goulburn in an attempt to establish a proper agreement between the

Government and himself. He suggested that the engine should be considered Government property, that professional engineers should investigate the costs of the work completed and assess future costs, the charges to be defrayed by the Government and that he should continue to direct the construction of the engine.

Goulburn rejected these proposals, and added in his reply that the Government were assisting 'an able and ingenious man of science, whose zeal had induced him to exceed the limits of prudence.'[17] He was under the total misapprehension that Babbage had begun the work on his own account, overspent himself, and then applied for Government assistance. Babbage replied vigorously to Lord Ashley, the intermediary in this correspondence, that such a charge was completely unfounded. He put forward as evidence the last paragraph of his letter to Davy; that the successful model of the machine seen by Goulburn and the Duke had been completed before his first interview with the Treasury in 1823; and that work had started on the present machine only as a consequence of this interview.[18]

The Government duly made a decision which was communicated to Babbage on 24 February 1830. This was a much more satisfactory arrangement that:

> 1st. Although the Government would not pledge themselves to COMPLETE the Machine, they were willing to declare it their property.
> 2nd. That professional Engineers should be appointed to examine the bills.
> 3rd. That the Government were willing to advance 3,000 l. more than the sum (6,000 l.) already granted.
> 4th. That, when the Machine was completed, the Government would be willing to attend to any claim of Mr. Babbage to remuneration, either by bringing it before the Treasury, or the House of Commons.[19]

This decision meant that a much more satisfactory way of financing the project had been agreed. In practice, the bulk of the work was performed by the engineer, one Clement, and when bills were sent in, they were forwarded by him to two other engineers, Donkin and Field, who examined their accuracy. When this was done, Babbage sent the bills to the Treasury and a warrant was issued directing payment to him. This often meant a time lag in Government payments, and in order to pay the workmen, Babbage frequently had to pay them himself before repayment from the Treasury.

In 1833, a small part of the engine was completed, capable of calculating tables with two or three orders of difference. This portion thoroughly justified confidence in the success of the whole project.

At the same time, it was agreed that the work should be transferred to fireproof premises in East Street, Manchester Square, adjacent to Babbage's own residence.

Babbage now found it no longer possible to make payments to his workmen in advance of money from the Treasury. On receiving this news, Clement immediately abandoned the work, dismissed the workmen and took with him all the tools, as he was apparently legally entitled to do. Most of this delicate equipment had been made specially for the construction of the Engine and progress was impossible without it. As a result, no further work was ever done on the difference engine, an additional £8000 having been spent under the revised financial agreement so that altogether £17 000 of Government money had been used on the scheme. When Babbage's own doubtless considerable sum of money expended on the engine is considered, the price may seem excessively high for the early nineteenth century, and might well have made the Governments following the year 1834 have second thoughts about continuing the project. However as Dionysius Lardner pointed out,[20] the improvements alone introduced into the steam engine by James Watt involved a capital expenditure of £50 000 between 1763 and 1782.

The situation, as far as Government support was concerned, now became more complicated in that Babbage, deprived of the means for further construction, had been thinking in more general terms, and worked out ways in which the analytical engine, a machine capable of performing any series of arithmetical operations and superseding by far the difference engine, might be built. To make the analytical engine viable, great technical improvements had to be made in performing arithmetical operations, especially those of addition and carriage. If such improvements could have been achieved, then they would also have revolutionised the operation of the difference engine, and probably rendered the one under construction obsolete before completion.

Babbage felt strongly that he ought to have a personal interview with the Head of the Government to inform him of this matter. On 26 September 1834, he requested an interview with Lord Melbourne. This was agreed to, then postponed, and soon after Lord Melbourne went out of office. He then requested an interview with the next Head of Government, the Duke of Wellington, specifically asking that a decision be reached concerning the difference engine, whether the Government wished that he should complete the task with Clement or with another engineer, or if he himself should be replaced as superintendent, or if the undertaking should be abandoned completely. Unfortunately, the Duke's term of office was also short.

On 15 May 1835, the difference engine was mentioned in the House of Commons when the Chancellor of the Exchequer stated distinctly that Babbage himself had not received a shilling for the whole project.

In subsequent correspondence, the Chancellor of the Exchequer (Spring Rice) seemed to form the wrong impression that Babbage was trying to persuade the Government to finance the analytical engine in place of the former one. Babbage insisted that he was only pressing for a decision regarding the completion of the difference engine alone.

The final decision, incredibly, was not made until November 1842, after eight years' total inactivity on the difference engine. After repeated attempts to recall the question to Sir Robert Peel's administration, he received the following letter from the unfortunate Goulburn, now Chancellor of the Exchequer again, dated 3 November 1842.

> The Solicitor-General has informed me that you are most anxious to have an early and decided answer as to the determination of the government with respect to the completion of your Calculating Engine. I accordingly took the earliest opportunity of communicating with Sir R. Peel on the subject.
>
> We both regret the necessity of abandoning the completion of a Machine on which so much scientific ingenuity and labour have been bestowed. But on the other hand, the expense which would be necessary in order to render it either satisfactory to yourself, or generally useful, appears on the lowest calculation so far to exceed what we should be justified in incurring, that we consider ourselves as having no other alternative.
>
> We trust that by withdrawing all claim on the part of the Government to the Machine as at present constructed, and by

placing it at your entire disposal, we may, to a degree, assist your
future exertions in the cause of science.
I am &c. Henry Goulburn.[21]

Babbage replied drily on 6 November:

> My Dear Sir,
> I beg to acknowledge the receipt of your letter of the 3rd of
> Nov., containing your own and Sir Robert Peel's decision
> respecting the Engine for calculating and printing mathemati-
> cal tables by means of Differences, the construction of which has
> been suspended about eight years.
> You inform me that both regret the necessity of abandoning
> the completion of the Engine, but that not feeling justified in
> incurring the large expense which it may probably require, you
> have no other alternative.
> You also offer, on the part of the Government, to withdraw all
> claim in the Machine as at present constructed, and to place it at
> my entire disposal, with the view of assisting my future exer-
> tions in the cause of science.
> The drawings and the parts of the Machine already executed
> are, as you are aware, the absolute property of the Government,
> and I have no claim whatever to them.
> Whilst I thank you for the feeling which that offer manifests, I
> must, under all the circumstances, decline accepting it.
> I am &c. C. Babbage.[22]

Babbage actually obtained an interview with Sir Robert Peel on
11 November. His feelings at the time are clearly brought out by
Nicolson. According to his own account of the interview, Babbage

> stated, that having given the original Invention to the Govern-
> ment – having superintended for them its construction – having
> demonstrated the possibility of the undertaking by the comple-
> tion of an important portion of it – and that the non-completion
> of the design arose neither from his fault nor his desire, but was
> the act of the Government itself, he felt that he had some claims
> on their consideration.
> He rested those claims upon the sacrifices he had made, both
> personal and pecuniary, in the advancement of the Mechanical
> Arts and of Science – on the anxiety and injury he had experi-
> enced by the delay of eight years in the decision of the Govern-
> ment on the subject, and on the great annoyance he had
> constantly been exposed to by the prevailing belief in the public
> mind that he had been amply remunerated by large grants of
> public money. Nothing, he observed, but some public act of the

Government could ever fully refute that opinion, or repair the injustice with which he had been treated.[23]

The outcome of this interview was entirely unsatisfactory, and Babbage left without attempting to propose some alternative plans he had prepared for the construction, with Government help, of both the difference and analytical engines.[24]

It certainly seems that the attitude of Peel was far less than gracious towards Babbage. According to A. De Morgan, reviewing Weld's *History of the Royal Society* for the *Athenaeum*, when the subject of the difference engine was raised in Parliament,

> Sir R. Peel turned if off with a joke in the House of Commons. He recommended that the machine should be set to calculate the time at which it would be of use. He ought rather to have advised that it should be set to compute the number of applications which remain unanswered before a Minister, if the subject were not one which might affect his parliamentary power.[25]

Babbage had every reason to feel aggrieved about his treatment by the successive Governments. They had failed to understand the immense possibilities of his work, ignored the advice of the most reputable scientists and engineers, procrastinated for eight years before reaching a decision about the difference engine, misunderstood his motives and the sacrifices he had made, and with the one exception in 1835, failed to protect him from public slander and ridicule. In an unsigned paper, 'Mr. Babbage's Calculating Engines', stated by De Morgan to be definitely by Babbage himself, it says: 'Neither the Science, nor the Institutions, nor the Government of his country have ever afforded him the slightest encouragement. When the Invention was noticed in the House of Commons, one single voice alone was raised in its favour.[26] The single voice, incidentally, was Hawes, the Member for Lambeth.

On the Government side, it must be stated that they did provide a total of £17 000 public funds for the project, and were prepared to count the project as a total loss and return the engine, parts and drawings to Babbage without cost, but this seems small consolation in view of the injustices he had received. It is interesting to note that Babbage contested Finsbury as a Radical in 1832 and 1834, but was bottom of the poll on both occasions.

The Royal Society do not seem to have offered any help, financial or practical, towards the construction, apart from their very favourable report for the Government. This was in distinct contrast to their counterparts in Sweden, but it might be that they were unable to forgive a man who had made such an attack on them in his *Reflections on the Decline of Science in England*, published in 1830.

The Government soon disposed of their unwanted property by presenting the working portion of the difference engine to the museum of King's College, London, in 1843. It was later transferred to the South Kensington Museum in 1872.

One of the ironies of the history of the difference engine is that in about 1834, a Stockholm printer named Scheutz learnt, through Lardner's article in the *Edinburgh Review*, of its construction, and began to build one himself. Working with his son and financed by the professors of the Academy of Sweden, with a small contribution from the Diet of Sweden, he completed a simpler machine than Babbage's, working on different mechanical principles for the arithmetic operations. This machine was unanimously awarded the Gold Medal at the Great Exposition of Paris in 1855. Babbage's attitude was very generous towards Scheutz, and he made a speech to the President and Fellows of the Royal Society after the delivery of the medals on 30 November 1855, describing the Swedish achievement and concluding that it was right the two Scheutzs should have been given the awards. The final irony was that in 1864 the British Government bought a copy of the Scheutz difference engine, and used it to compute a set of life tables related to mathematical principles worked out by Babbage nearly forty years earlier!

W. Farr, giving his inaugural address as President of the Statistical Society in 1871, recalled that in 1855:

> The English Life Table No. 3 was then under consideration at the General Register Office, and perceiving that, by employing logarithms and properly adjusting the formulas, the whole of the tables for joint lives in all combinations, as well as single lives and discounted values of future payments, could be computed by the method of finite differences, I brought the matter under the notice of the Registrar-General, who submitted to Her Majesty's Treasury that the idea of a difference engine originated in England; that it had now been realised, that such an engine was capable of performing useful work, and that a replica of the Swedish engine, with improvements, should be constructed by the well-known firm of Bryan Donkin & Co. who

offered to complete it for £1,200. The Astronomer-Royal concurred in the recommendation. This machine at the General Register Office calculated and stereoglyphed, under my supervision, the fundamental columns of the English Life Table, which, with all necessary explanations, was published in 1864.[27]

After the Government refusal to give any more financial assistance, Babbage worked for some years on the theoretical principles and drawings for the analytical engine. Having mastered these in 1848 he decided to use the improvements and simplifications he had devised to construct a new set of drawings for a second difference engine.

In 1852, the Earl of Rosse who was well acquainted with these drawings persuaded a not surprisingly reluctant Babbage to resubmit a case to the Government for aid in the construction of difference engine No. 2. He also suggested that, to make the project more practical, the Government should apply to the President of the Institution of Civil Engineers to ascertain

> 1st. Whether it was possible, from the drawings and notations, to make an estimate of the cost of constructing the machine? 2ndly. In case this question was answered in the affirmative – then, could a Mechanical Engineer be found who would undertake to construct it, and at what expense?[28]

These two very sensible suggestions, if answered in the affirmative, would have resolved the major difficulties of the first difference engine, namely the indefinite cost and the personal demands made on Babbage.

Lord Rosse conveyed his plan, together with a rather pathetic letter by Babbage to the Premier, Lord Derby, on 8 June 1852.[29] In the letter, Babbage had outlined the unfortunate history of the first engine and given a brief account of his new improvements through his thinking on the analytical engine. One point of interest is that Babbage states[30] that in addition to the £17 000 supplied from public funds, his own expenditure on this and other scientific works was in the order of £20 000.

The request was passed by Lord Derby to his Chancellor of the Exchequer, Benjamin Disraeli, whose reply was even worse than Goulburn's. An extract states that 'Mr. Babbage's projects appear to be so indefinitely expensive, the ultimate success so problematical, and the expenditure certainly so large and so utterly incapable of being calculated, that the Government would not be justified in taking upon itself any further liability.'[31]

Babbage bitterly complained that the Earl of Rosse's suggestion had not even been considered. The Chancellor had refused to consult expert advice, but taken it upon himself to pronounce on the cost of the project. He had lightly dismissed as problematical the possibility of ultimate success of the machine, when both the mathematical principle and the mechanical working as exemplified in the completed portion of the first difference engine were known by all to be perfectly sound. Babbage said further that there was not a single reputable mathematician who would not vouch for the usefulness of such a machine.

The Earl of Rosse, in a Presidential address to the members of the Royal Society at their anniversary on 30 November 1854, said:

> That the first great effort to employ the powers of calculating mechanism, in aid of the human intellect, should have been suffered in this great country to expire fruitless, because there was no tangible evidence of immediate profit, as a British subject I deeply regret, and as a Fellow my regret is accompanied with feelings of bitter disappointment. Where a question has once been disposed of, succeeding Governments rarely reopen it, still I thought I should not be doing my duty if I did not take some opportunity of bringing the facts once more before Government. Circumstances had changed, mechanical engineering had made much progress; the tools required and trained workmen were to be found in the workshops of the leading mechanists, the founder's art was so advanced that casting had been substituted for cutting, in making the change wheels, even of screw-cutting engines, and therefore it was very probable that persons would be found willing to undertake to complete the Difference Engine for a specific sum.
>
> That finished, the question would then have arisen, how far it was advisable to endeavour, by the same means, to turn to account the great labour which had been expended under the guidance of inventive powers the most original, controlled by mathematics of a very high order; and which had been wholly devoted for so many years to the great task of carrying the powers of calculating machinery to its utmost limits. Before I took any step I wrote to several very eminent men of science, inquiring whether, in their opinion, any great scientific object would be gained if Mr. Babbage's views, as explained in Menabrea's little essay, were completely realized. The answers I received were strongly in the affirmative. As it was necessary the

subject should be laid before the Government in a form as practical as possible, I wrote to one of our most eminent mechanical engineers to enquire whether I should be safe in stating to Government that the expense of the Calculating Engine had been more than repaid in the improvements in mechanism directly referable to it; he replied – unquestionably.[32]

This was one of the few tributes that Babbage received in his lifetime, and indicates clearly some of the indirect benefits that mechanical engineering had received from the uncompleted difference engine.

The melancholy story of Babbage's attempts to obtain proper State support for his computers is an indictment on the successive governments that ruled the nation at this time. When the Industrial Revolution was nearing its peak the Government showed itself to be not particularly interested in the advance of science or technology. In making important decisions of policy on these matters the advice of such learned bodies as the Royal Society and the Institute of Civil Engineers was ignored in favour of politicians' hunches. The only Head of Government to have emerged with any credit from this episode was the Duke of Wellington and even he on one occasion preferred his private judgement on a scientific question to that of a considered report by some of the most eminent members of the Royal Society.

Babbage made no further communications to the Government on this subject.

THE ANALYTICAL ENGINE

The history of the analytical engine is necessarily briefer than that of the difference engine. There is no mathematical principle to explain, only a logical arrangement; no dispute over finance, as Babbage determined from his previous experience to bear the entire cost himself. As his mother said to him: 'You have advanced far in the accomplishment of a great object, which is worthy of your ambition. You are capable of completing it. My advice is – pursue it, even if it should oblige you to live on bread and cheese.'[33] Babbage agreed completely with this maternal advice. He financed the whole project, which lasted nearly forty years, but without, it would appear, his ever having to live as frugally as suggested.

Despite all his efforts, the analytical engine was never anywhere near completed, and belongs properly to the museum of magnificent ideas that failed to materialise. The conception of this engine was, however, far superior to that of the difference engine, and, as will be shown, compares very well with that of a modern computer.

As with the difference engine, Babbage published no systematic account of the principle and workings of the analytical engine. His only published work on the project occupies a single chapter in each of his books *Passages from the Life of a Philosopher* and *The Exposition of 1851*, London, 1851. More detailed accounts of the engine are given in L. F. Menabrea's *Sketch of the Analytical Engine*, translated by Lady Lovelace, London, 1843, and H. P. Babbage's address to the British Association in 1888, published in the latter's *Babbage's Calculating Engines*. As Babbage himself thoroughly approved of Menabrea's sketch, and his son worked closely with him on the analytical engine, we may take these two accounts as authoritative. There is also the article published anonymously in the *Philosophical Magazine* for September 1843, entitled 'Mr. Babbage's calculating engines'. In this, the reason for Babbage's unwillingness to publish a comprehensive account of either machine is stated: 'The circumstance of his not having printed a description of either Engine has arisen entirely from his determination never to employ his mind upon the *description* of those Machines so long as a single difficulty remained which might *limit* the power of the Analytical Engine.'[26]

The first question to be considered is 'how and when did Babbage first conceive the principle for the Analytical Engine'? In chapter VIII, headed 'Of the analytical engine', of *Passages from the Life of a Philosopher*, he begins: 'The circular arrangement of the axes of the Difference Engine round large central wheels led to the most extended prospects. The whole of arithmetic now appeared within the grasp of mechanism. A vague glimpse of an Analytical Engine at length opened out, and I pursued with enthusiasm the shadowy vision.'[34] Unfortunately, there is nothing in the context to indicate precisely when this idea occurred. H. P. Babbage is much more specific about this:

> 4. The idea of the Analytical Engine arose thus: When the fragment of the Difference Engine, now in the South

Kensington Museum, was put together early in 1833, it was
found that, as had been before anticipated, it possessed powers
beyond those for which it was intended, and some of them could
be and were demonstrated on that fragment.

5. It was evident that by interposing a few connecting-wheels,
the column of Result can be made to influence the last Difference, or other part of the machine in several ways. Following
out this train of thought, he first proposed to arrange the axes of
the Difference Engine circularly, so that the Result column
should be near that of the last Difference, and thus easily within
reach of it. He called this arrangement, 'the engine eating its
own tail'. But this soon led to the idea of controlling the machine
by entirely independent means, and making it perform not only
Addition, but all the processes of arithmetic at will in any order
and as many times as might be required. Work on the Difference Engine was stopped on 10th April, 1833, and the first
drawing of the Analytical Engine is dated in September, 1834.[35]

It seems from these two accounts that the idea of the analytical
engine arose from the facility of the difference engine to calculate
series automatically if arranged mechanically in a suitable way.
The sort of series that could thus be obtained seem undoubtedly
to be those in which, for example, the first or second difference is
made equal to the units figure of the previous result, and if this is
the case, then Babbage certainly had the idea long before the
suggested 1833. As I have already mentioned, Babbage published two papers on this property of the difference engine as
early as 1822 and mentioned in his letter to Sir Humphry Davy
the possibility of performing such calculations, with the hint of
more advanced machines to come.

It may be concluded that Babbage certainly had a variety of
ideas germinating in his mind at all stages of the construction and
most probably had the thought of the analytical engine very
shortly after his first work on the difference engine. These ideas
were worked out more thoroughly in 1833 when he was, through
Clement's defection, first deprived of the means of continuing
the difference engine, and, as stated by his son, the first drawing
for the analytical engine appeared in September 1834.

The general principle of the working of this engine is given
most clearly in the chapter from the *Passages from the Life of a
Philosopher*. In its logical design, the engine contained five essen-

Plate 2. The difference engine

tial parts, which may be named the store, mill, control, input and output. To quote Babbage's account of the first two of these:

> 1st. The store in which all the variables to be operated upon, as well as all those quantities which had arisen from the result of other operations, are placed.
> 2nd. The mill into which the quantities about to be operated upon are always brought.[36]

The store, as its name suggests, was the part of the machinery in which all the numbers to be used in the particular calculation were brought, together with the new numbers that had arisen in the course of the computation. The size of the store could be chosen arbitrarily, and Babbage had in mind a very large one, capable of a thousand numbers each to fifty significant figures.

The actual arithmetical operations were to be performed one at a time, in the mill. On receiving the appropriate instruction, the required two numbers would be brought from the store into the mill, operated upon, if necessary returned to their places in the store, and the result of the operation recorded elsewhere in the store.

The control of the sequence of operations was by means of a Jacquard loom. This was the only device capable of being adapted to the purpose of the analytical engine at the time, and Babbage was swift to appreciate its possibilities. In 1801, J. M. Jacquard of Lyons made the first successful automatic drawloom by means of a series of instructions given to the threads by punched cards. Thousands of Jacquard looms were made in the early part of the nineteenth century, and Babbage possessed the most complicated of these, a self-portrait of Jacquard woven by means of 24 000 cards, each one capable of receiving 1050 punch-holes. This system was ideal for controlling the sequence of simple arithmetical instructions required in a calculation by the analytical engine.

Consequently the input to the engine was a set of punched cards. Babbage used two types of card for determining the operation, and locating the numbers to be operated on: 'There are therefore two sets of cards, the first to direct the nature of the operations to be performed – these are called operation cards: the other to direct the particular variables on which those cards are required to operate – these latter are called variable cards.'[37]

Each calculation, then, consisted of a string of punched cards in the correct order for the operations and the variables. The way in which the programme was written is fully described in Menabrea's memoir, which we will consider shortly.

Finally, the output of the engine was intended to be presented in an equally ambitious fashion:

> The Analytical Engine will contain,
> 1st. Apparatus for printing on paper, one, or, if required, two copies of its results.
> 2nd. Means for producing a stereotype mould of the tables or results it computes.
> 3rd. Mechanism for punching on blank pasteboard cards or metal plates the numerical results of any of its computations.[38]

These five parts are the major logical components of any modern computer.

One great advantage of the engine, which Babbage foresaw, was the possibility of a library of programmes:

> Every set of cards made for any formula will at any future
> time recalculate that formula with whatever constants may be
> required.
> Thus the Analytical Engine will possess a library of its own.
> Every set of cards once made will at any future time reproduce
> the calculations for which it was first arranged. The numerical
> value of its constants may then be inserted.[39]

The speed of the machine would, of course, have been very
slow by modern standards, but, according to Babbage's estimate,
sufficiently faster than a human calculator to justify its construction:

> Supposing the velocity of the moving parts of the Engine to be
> not greater than forty feet per minute, I have no doubt that
> Sixty additions or subtractions may be completed and printed
> in one minute.
> One multiplication of two numbers, each of fifty figures, in
> one minute.
> One division of a number having 100 places of figures by
> another of 50 in one minute.[40]

In view of the mechanical limitations imposed on the speed of
the engine it was necessary for Babbage to ensure that the basic
calculations were performed as quickly as possible. His greatest
problem was that of carriage in addition, and the solution to this
produced one of the most ingenious devices in the whole mechanism. Before describing this solution, it will be necessary to show by
examples the nature of the difficulty.

Addition was performed basically by bringing together the
corresponding successive digits of the two numbers, then adding
by means of giving each wheel divided into ten units the appropriate number of turns simultaneously, and then performing the
carriage as a secondary operation. To add together, for example,
4953 and 2308, the corresponding digits are placed together,

4953

2308

the 3 in the units column of the first number receives 8 additional
increments, so that it turns eventually to the figure 1, and as the 9
passed to zero, a figure 1 for carrying automatically goes to the

next stage. This process takes place simultaneously with all columns so that the first stage of the calculation may be expressed as:

```
4 9 5 3
2 3 0 8
———————
6 2 5 1
1   1
```

The carriage figures are then added, so that the whole calculation may be performed in two stages as:

```
4 9 5 3
2 3 0 8
———————
6 2 5 1
1   1
———————
7 2 6 1
```

As far as time is concerned, the first stage may be considered as having taken a maximum of nine units, since nine is the largest amount by which any wheel may have turned, and the carriage in one unit.

If, however, it is required to add 4953 and 2348, the problem requires three stages as:

```
4 9 5 3
2 3 4 8
———————
6 2 9 1
1   1
———————
7 2 0 1
    1
———————
7 3 0 1
```

A more critical addition is

```
3 9 5 3
6 0 4 8
─────────
9 9 9 1
      1
─────────
9 9 0 1
    1
─────────
9 0 0 1
  1
─────────
0 0 0 1
1
─────────
1 0 0 0 1
```

needing five stages.

It can easily be seen that when fifty digit numbers are added, a maximum of fifty successive carriages is possible. In this case, the addition would take not ten but fifty-nine units of time to perform, and Babbage was, therefore, most anxious to solve this time-consuming question of successive carriage. He states concerning this:

> The most important part of the Analytical Engine was undoubtedly the mechanical method of carrying the tens. On this I laboured incessantly, each succeeding improvement advancing me a step or two. The difficulty did not consist so much in the more or less complexity of the contrivance as in the reduction of the *time* required to effect the carriage. Twenty or thirty different plans and modifications had been drawn.[41]

Although Babbage eventually solved this problem, his account in the autobiography expressed only his delight in the achievement and gives no indication of the method. Fortunately, the device for anticipatory carriage is described in great detail in H. P. Babbage's account, together with a description of the mechanical way in which a straight-forward carriage is performed.

There is a series of blocks two for each place, the lower block is made to serve for the two events. The upper one has a projecting arm, which when moved circularly engages a toothed wheel and moves it on one tooth affecting the figure disc similarly. It rests on the lower block which moves it up and down with itself always. After the addition is completed, should a carriage have become due and the warning lever have been pushed aside, it is made (by motion from the main shaft) to actuate the lower block and throw it into 'Chain', when this is raised (again by motion from the main shaft) it raises the upper block which thus affects ordinary carriage; but supposing no carriage to have become due, there will be at the window either a 9 or some less number. The latter case may be dismissed at once; as a carriage arriving to it will cause no carriage to be passed on. Should there be a 9 at the window the warning-lever cannot have been pushed aside, but in every place where there is a 9 another lever, again by motion from the main shaft comes into play and pushes the lower block into 'Chain'; not, of course, into 'Chain' for ordinary carriage, but into another position for Chain for 9's, so that should there come a carriage from below, the Chain for 9's will be raised as far as it extends and effect the carriages necessary, be it for a single place, or for several, or for many. Should there be no carriage from below to disturb it the Chain remains passive; it has been made ready for a possible event which has not occurred. A certain time is required for the preparation of 'Chain', but that done all the carriages are then effected simultaneously. A piece of mechanism for Anticipating Carriage on this plan to twenty-nine figures exists and works perfectly.[42]

If, then, a string of 9s occurs after the first addition, a single carriage will perform all the successive carriages simultaneously. Otherwise the 9s remain unchanged. This 'all-or-none' device is remarkably similar in principle to the 'and' box used extensively in modern computers.

Babbage goes on to discuss in the *Passages* how various fundamental difficulties of an arithmetical or algebraic nature may be overcome. These problems he enumerates as follows: relating to arithmetic,

(a) The number of digits in *each constant* inserted in the Engine must be without limit.

(b) The number of constants to be inserted in the Engine must also be without limit.

(c) The number of operations necessary for arithmetic is only four, but these four may be repeated an *unlimited* number of times.

(d) These operations may occur in any order, or follow an *unlimited* number of laws.

The following conditions relate to the algebraic portion of the Analytical Engine:–

(e) The number of *litteral* constants must be *unlimited*.

(f) The number of *variables* must be *without limit*.

(g) The combinations of the algebraic signs must be *unlimited*.

(h) The number of *functions* to be employed must be *without limit*.[43]

These difficulties all arise from the obvious fact that any such engine has only a finite storage space. All of them can be solved by making the store sufficiently large, by transposing problems concerning an infinity of space into those of an infinity of time, and by increasing indefinitely the number of punched cards.

It has already been said that Babbage envisaged a store consisting of a thousand numbers, each to fifty significant figures, and this he very reasonably considered more than large enough to deal with any practical problem. However, if it is required, for example, to multiply together two numbers each consisting of a hundred figures, the problem can be solved on the engine by considering the two numbers as $a \cdot 10^{50} + b$ and $a^1 \cdot 10^{50} + b^1$, where a, b, a^1, b^1 all have less than fifty digits. The product of two such numbers will be $aa^1 10^{100} + (ab^1 + a^1 b)10^{50} + bb^1$, containing four pairs of factors $aa^1, a^1 b^1, ab^1, bb^1$ each factor of which has less than fifty digits. The multiplication can, therefore, be performed in the engine, the time required being roughly four times that for performing an ordinary multiplication, and twice the storage space is required. In general, if it is required to multiply numbers with k times the number of digits available in the engine, the time taken will be roughly k^2 times the normal. At the same time, an engine capable of storing m constants will be reduced to a storage of m/k.

This indicates how objection (a) can be overcome, but the solution leads directly to objection (b) which concerns the number of constants that can be stored in the engine. If it can be shown that in effect an unlimited number of constants can be stored, then problems (a) and (b) are immediately surmounted. Babbage achieved this by manipulating the input and output devices. All numbers required, and those arising in the course of the computation, may be recorded by the printing device, and therefore returned to the computation by means of the punched card system whenever required. Thus, although the engine can only contain a finite number of numbers at a time, any combination of the numbers required in a problem can be put into it at any desired time.

Objections (c) and (d), concerning the unlimited number of operations to be performed in a complicated programme, can easily be overcome by introducing a sufficient number of punched cards. Having clearly answered the four arithmetical problems, Babbage then proceeds to show that the algebraic difficulties are almost equivalent.

A further device enabled the engine to repeat a given sequence of instructions when required, and to make its own act of judgement about when to do this. Babbage describes a meeting he had with some Italian philosophers at Turin in 1840. In the course of these discussions, M. Mosotti mentioned to him that 'he had no conception how the machine could perform the act of judgment sometimes required during an analytical inquiry, when two or more different courses presented themselves, especially as the proper course to be adopted could not be known in many cases until all the previous portion had been gone through.'[44] Babbage answered this question by describing how a problem on the location of roots of an equation of any degree might be solved. In this type of problem it would be necessary to teach the engine to know when to change from one set of cards to another, and back again repeatedly, at intervals not known to the person conducting the computation. The decision is achieved by ascertaining at a certain point in the calculation whether a given number becomes positive or negative, and the result of this test can be used to govern the next set of instructions. 'Mechanical means have been provided for backing or advancing the operation cards to any extent. There exist means of expressing the conditions under which these various processes are required to be called into play.'[45]

It appears then that Babbage intended his analytical engine to be able to make decisions in the course of a calculation, and perform 'looped' instructions as many times as required.

During the course of this meeting in Turin, L. F. Menabrea, later to become a general in Garibaldi's army, and Prime Minister of Italy, collected the material for his description of the analytical

Plate 3. The analytical engine

engine first published in the Bibliothèque Universalle de Genève, *tome* xli, October 1842.

This short article was translated into English by the Countess of Lovelace, Ada Augusta King, the only child of Lord Byron. At Babbage's suggestion, she wrote notes on Menabrea's article which turned out to be three times as long as the memoir itself. Babbage, who had a high regard for her mathematical ability, said:

> We discussed together the various illustrations that might be introduced: I suggested several, but the selection was entirely her own. So also was the algebraic working out of the different problems, except, indeed, that relating to the numbers of Bernoulli, which I had offered to do to save Lady Lovelace the trouble. This she sent back to me for an amendment, having detected a grave mistake which I had made in the process ... These two memoirs taken together furnish, to those who are capable of understanding the reasoning a complete demonstration – *That the whole of the developments and operations of analysis are now capable of being executed by machinery.*'[46]

Both Menabrea and Lady Lovelace emphasise the dual nature of Babbage's invention and praise his creative skill in both the mathematical and mechanical aspects. Their intention is clearly to say as little as possible about the actual construction of the analytical engine and to concentrate on its analytical nature. The result is that there is little to add to what has already been said in this chapter apart from their most interesting discussion of the way in which the calculation is presented to the engine, which is equivalent to the first recorded example of programming for a computer.

The translation and notes by Lady Lovelace appeared in the *Scientific Memoirs*, 1843, **3**, 666–731, printed by Richard and John E. Taylor.

After discussing the mathematical principle of the difference engine as a complete contrast to that of the analytical engine, Menabrea proceeds to show how a calculation might actually be performed on the latter. He represents the columns on which numbers are recorded in the store by the symbols V_0, V_1, V_2, V_3, etc. and explains how these are brought two at a time into the mill and the result of the operation recorded elsewhere.

Suppose it is required to solve the simultaneous equations

$$mx + ny = d$$
$$m^1 x + n^1 y = d^1$$

with solution

$$x = \frac{dn^1 - d^1 n}{n^1 m - nm^1}$$

and an analogous expression for y. Then to evaluate x using the engine, first locate the six numbers m, n, d, m^1, n^1, d^1 in the locations V_0, V_1, V_2, V_3, V_4, V_5, respectively. Then the series of operations to evaluate x may be tabulated as follows:[47]

| | Operation-cards | Cards of the variables | | |
Number of the operations	Symbols indicating the nature of the operations	Columns on which operations performed	Columns which receive results of operations	Progress of the operations
1	$\times$	$V_2 \times V_4 =$	V_8	$= dn^1$
2	$\times$	$V_5 \times V_1 =$	V_9	$= d^1 n$
3	$\times$	$V_4 \times V_0 =$	V_{10}	$= n^1 m$
4	$\times$	$V_1 \times V_3 =$	V_{11}	$= nm^1$
5	$-$	$V_8 - V_9 =$	V_{12}	$= dn^1 - d^1 n$
6	$-$	$V_{10} - V_{11} =$	V_{13}	$= n^1 m - nm^1$
7	$\div$	$\dfrac{V_{12}}{V_{13}} =$	V_{14}	$= x = \dfrac{dn^1 - d^1 n}{n^1 m - nm^1}$

It is fairly clear from this example that in each operation two numbers from the required locations are brought into the mill and the result given a different address, which can be used in further operations.

Lady Lovelace has a slightly more elaborate way of writing out this sequence of operations. She points out that when a particular variable column is brought into the mill, it may either retain its numerical value, have this value changed to zero if no further use is needed for it, or have the result of the operation recorded on it.

She retains Menabrea's notation $V_0, V_1, V_2, \ldots$, as far as lower suffices are concerned, to denote location, but also introduced higher suffices $^0V, ^1V, ^2V, \ldots$ to denote the state of the variable column. Any column which is given a number is labelled ^1V, and if this number is altered during the operations it becomes successively $^2V, ^3V, \ldots$. But if the number is discarded and the column reduced to zero, ^0V is recorded. As it is tidier to reduce all columns, except the results columns, to zero in the course of the calculation, she tabulates the calculation as on facing page.

Thus the results are obtained after eleven operations, and each column except for the last two is brought back to zero in the course of the calculation. Lady Lovelace justifies her upper suffix notation by remarking that if it is desired to replace a number in a certain column by the result of an operation involving itself in the mill, it is better to record the event as $^{m+1}V_n = {}^mV_p + {}^mV_n$, rather than the confusing $V_n = V_p + V_n$.

Menabrea also describes how differences in sign are dealt with. At the top of each column is a disc, similar to the discs containing the successive digits of the given number. According to whether this digit is even or uneven, the number below is considered positive or negative, respectively. When two numbers are added, if the signs are both even or both odd, one of the numbers is transferred to the mill so that the resulting number has the same sign as the two original. If one is even and the other odd, the mechanism automatically changes the operation to a subtraction and the result is given the sign of the longer of these two numbers. For multiplication, the two numbers indicating the signs are added so that like signs give a positive result and unlike ones negative. Similarly, for division, the two numbers are subtracted.

Menabrea admits that on this topic he is only surmising, not having discussed the matter with Babbage, but Lady Lovelace comments that this, remarkably, is exactly what the inventor intended.

Another interesting device described by Menabrea concerns repeated instructions to the engine. If it is required to calculate ab^n, then a, b, n are placed in V_0, V_1, V_2, respectively, and the result recorded in V_3. To compute b^n, there is a certain registering apparatus in the machine which alters n into $n-1$ when b is first multiplied by b. At the second multiplication $n-1$ effaces a unit to $n-2$, and the engine continues to multiply b by itself until the registering apparatus marks zero.

No. of operations	Nature of operations	Variables for data						Working variables									Variables for results	
		$^1V_0$	$^1V_1$	$^1V_2$	$^1V_3$	$^1V_4$	$^1V_5$	$^0V_6$	$^0V_7$	$^0V_8$	$^0V_9$	$^0V_{10}$	$^0V_{11}$	$^0V_{12}$	$^0V_{13}$	$^0V_{14}$	$^0V_{15}$	$^0V_{16}$
		m	n	d	m^1	n^1	d^1											
1	×	m				n^1		mn^1										
2	×		n		m^1				nm^1									
3	×			d						dn^1								
4	×		0				d^1				d^1n							
5	×	0					0					d^1m						
6	×			0	0								dm^1					
7	−							0	0					mn^1-m^1n				
8	−									0	0				dn^1-d^1n			
9	−											0	0			d^1m-dm^1		
10	÷													mn^1-m^1n	0		$\frac{dn^1-d^1n}{mn^1-m^1n}=x$	
11	÷													0		0		$\frac{d^1m-dm^1}{mn^1-m^1n}=y$

Again, if two polynomials are multiplied together, the engine has a facility for collecting up like terms. For if two of the terms are ax^{m+p} and bx^{n+q}, the engine will subtract $n+q$ from $m+p$ and if, and only if, the result is zero, it will then add a to b.

Most of Lady Lovelace's notes are elaborations on the points made by Menabrea, together with some complicated programmes of her own, the most complex of these being one to calculate the sequence of Bernoulli numbers.

Babbage left a further account of the analytical engine in his book *The Exposition of 1851*, chapter XIII. This chapter includes a description of his method for achieving simultaneous carriage, which is less detailed than H. P. Babbage's account, which we have already considered. Of greater interest is his statement concerning the state of the drawings for the engine at that time, and his sense of frustration about the whole project:

> The drawings of the *Analytical Engine* have been made entirely at *my own cost*: I instituted a long series of experiments for the purpose of reducing the expense of its construction to limits which might be within the means I could myself afford to supply. I am now resigned to the necessity of abstaining from its construction, and feel indisposed even to finish the drawings of one of its many general plans. As a slight idea of the state of the drawings may be interesting to some of my readers, I shall refer to a few of the great divisions of the subject.
>
> ARITHMETICAL ADDITION: – About a dozen plans of different mechanical movements have been drawn. The last is of the very simplest order.
>
> CARRIAGE OF TENS: – A large number of drawings have been made of modes of carrying tens. They form two classes, in one of which the carriage takes place successively: in the other it occurs simultaneously, as will be more fully explained at the end of this chapter.
>
> MULTIPLYING BY TENS: – This is a very important process, though not difficult to contrive. Three modes are drawn; the difficulties are chiefly those of construction, and the most recent experiments now enable me to use the simplest form.
>
> DIGIT COUNTING APPARATUS: – It is necessary that the machine should count the digits of the numbers it multiplies and divides, and that it should combine these properly with the number of decimals used. This is by no means as easy as the former operation: two or three systems of contrivances have been drawn.

COUNTING APPARATUS: – This is an apparatus of a much more general order, for treating the indices of functions and for the determination of the repetitions and movements of the Jacquard cards, on which the Algebraic developments of functions depend. Two or three such mechanisms have been drawn.

SELECTORS: – The object of the system of contrivances thus named is to choose in the operation of Arithmetical division the proper multiple to be subtracted; this is one of the most difficult parts of the engine, and several different plans have been drawn. The one at last adopted is, considering the subject, tolerably simple. Although division is an inverse operation, it is possible to perform it entirely by mechanism without any tentative process.

REGISTERING APPARATUS: – This is necessary in division to record the quotient as it arises. It is simple, and different plans have been drawn.

ALGEBRAIC SIGNS: – The means of combining these are very simple, and have been drawn.

PASSAGE THROUGH ZERO AND INFINITY: – This is one of the most important parts of the Engine, since it may lead to a totally different action upon the formulae employed. The mechanism is much simpler than might have been expected, and is drawn and fully explained by notations.

BARRELS AND DRUMS: – These are contrivances for grouping together certain mechanical actions often required; they are occasionally under the direction of the cards; sometimes they guide themselves, and sometimes their own guidance is interfered with by the Zero Apparatus.

GROUPINGS: – These are drawings of several of the contrivances before described, united together in various forms. Many drawings of them exist.

GENERAL PLANS: – Drawings of all the parts necessary for the Analytical Engine have been made in many forms. No less than thirty different general plans for connecting them together have been devised and partially drawn; one or two are far advanced. No. 25 was lithographed at Paris in 1840. These have been superseded by simpler or more powerful combinations, and the last and most simple has only been sketched.

A large number of Mechanical Notations exist, showing the movements of these several parts, and also explaining the processes of arithmetic and algebra to which they relate. One amongst them, for the process of division, covers nearly thirty large folio sheets.[48]

These ten basic contrivances grouped together in various combinations seem adequate to perform any sequence of arithmetical operations, and had it been possible to construct them according to Babbage's directions, there is little doubt that they would have formed the first automatic computer as we understand the term. Babbage's disappointment with the progress of the engine and his threat to abandon work on the project in 1851 did not materialise, for his recorded drawings date until at least 1864. It is quite probable that he continued work until near his death in 1871, and even in his declining years it cannot be doubted that he was still making improvements to the design. His ideas were too far ahead of his contemporaries for anyone to realise the importance of what he was doing, and to give him any reasonable assistance. As a description of a possible direction taken by Babbage in his declining years, the anecdote recalled by Lord Moulton[49] may be taken either as an accurate account or an illustration of Babbage's somewhat elaborate sense of humour. Babbage's own feelings about the analytical engine are given in one of the last chapters of his *Passages from the Life of a Philosopher*:

> If I survive some few years longer, the Analytical Engine will exist, and its works will afterwards be spread over the world. If it is the will of that Being, who gave me the endowments which led to that discovery, that I should not survive to complete my work, I bow to that decision with intense gratitude for those gifts: conscious that through life I have never hesitated to make the severest sacrifices of fortune, and even of feelings, in order to accomplish my imagined mission.
>
> The great principles on which the Analytical Engine rests have been examined, admitted, recorded, and demonstrated. The mechanism has now been reduced to unexpected simplicity. Half a century may probably elapse before any one without those aids which I leave behind me, will attempt so unpromising a task. If, unwarned by my example, any man shall undertake and shall succeed in really constructing an engine embodying in itself the whole of the executive department of mathematical analysis upon different principles or by simpler mechanical means, I have no fear of leaving my reputation in his charge, for he alone will be fully able to appreciate the nature of my efforts and the value of their results.[50]

Babbage's estimate of fifty years was too generous to his contemporaries and successors. It was over seventy years after his prediction was made that the first working automatic computer was produced. Certainly his own age was not prepared to agree about the feasibility of such a machine, nor to agree with its inventor's optimism about the simplicity of the devices. Seven years after his death, the British Association appointed a committee to enquire into the possibility of completing his work. They reported as follows:

> Report of the Committee consisting of Professor Cayley, Dr. Farr, J. W. L. Glaisher, Dr. Pole, Professor Fuller, Professor A. B. W. Kennedy, Professor Clifford and Mr. C. W. Merrifield, appointed to consider the advisability and to estimate the expense of constructing Mr. Babbage's Analytical Machine, and of printing Tables by its means.[51]

> *General Conclusions and Recommendation*
> 1. We are of opinion that the labours of Mr. Babbage, firstly on his Difference Engine, and secondly on his Analytical Engine, are a marvel of mechanical ingenuity and resource.
> 2. We entertain no doubt as to the utility of such an engine as was in his contemplation when he undertook the invention of his analytical engine, supposing it to be successfully constructed and maintained in efficiency.
> 3. We do not consider that the possibilities of its misuse are any serious drawback to its use or value.
> 4. Apart from the question of its saving labour in operations now possible, we think the existence of such an instrument would replace within reach much which, if not actually impossible, has been too close to the limits of human skill and endurance to be practically available.
> 5. We have come to the conclusion that in the present state of the design of the engine, it is not possible for us to form any reasonable estimate of its cost, or of its strength and durability.
> 6. We are also of the opinion that, in the present state of the design, it is not more than a theoretical possibility; that is to say, we do not consider it a certainty that it could be constructed and put together so as to run smoothly and correctly, and to do the work expected of it.
> 7. We think that there remains much detail to be worked out, and possibly some further invention needed, before the design can be brought into a state in which it would be possible to judge whether it would really so work.
> 8. We think that a further cost would have to be incurred in order to bring the design to this stage, and that it is just possible that a mechanical failure might cause the expenditure to be lost.

9. While we are unable to frame any exact estimates, we have reason to think that the cost of the engine, after the drawings are completed, would be expressed in tens of thousands of pounds at least.

10. We think there is even less possibility of forming an opinion as to its strength and durability than as to its feasibility or cost.

11. Having regard to all these considerations, we have come, not without reluctance, to the conclusion, that we cannot advise the British Association to take any steps, either by way of recommendation or otherwise, to procure the construction of Mr. Babbage's Analytical Engine and the printing tables by its means.

12. We think it, however, a question for further consideration whether some specialized modification of the engine might not be worth construction, to serve as a simple multiplying machine, and another modification of it arranged for the calculation of determinants, so as to serve for the solution of simultaneous equations. This, however, inasmuch as it involves a departure from the general idea of the inventor, we regard as lying outside the terms of reference, and therefore perhaps rather for the consideration of Mr. Babbage's representatives than ours. We accordingly confine ourselves to the mere mention of it by way of suggestion.[52]

CONCLUSION

It only remains to give an assessment of the importance of Babbage's work in the history of computation, and this is quite a simple task.

Briefly, he began his project at a time when the only calculating machines available were crude hand-operated adders and multipliers, which usually did not work, and he produced, almost alone, the major ideas upon which all modern computers are constructed. Even in the twentieth century with all the encouragement, finances, resources and techniques that Babbage never had, no-one attempted to devise such a machine until as late as 1937. It is almost banal to say that Babbage's work was not in any way fully appreciated by his contemporaries, much less the authorities of his day, but it still seems extraordinary that having demonstrated a working model of a difference engine in 1822, he should have had to waste twenty years of creativity personally supervising the construction of a larger type working on the principle, only to see the whole project abandoned. For the more important analytical engine we have seen that Babbage asked for no assistance and certainly received none. It is tantalising to compare the support given to Babbage with that supplied to the first computer of this century:

> The machine now described 'The Automatic Sequence Controlled Calculator', is a realisation of Babbage's project in principle, although its physical form has the benefit of twentieth century engineering and mass-production methods.
>
> Prof. Howard H. Aiken (also Commander U.S.N.R.) of Harvard University inspired the International Business Machines Corporation (I.B.M.) to collaborate with him in constructing a new machine, largely composed of standard Hollerith counters, but with a superimposed and specially designed tape sequence control for directing the operations of the machine. The foremost I.B.M. engineers were assigned to the task; many of their new inventions are incorporated as basic units.[53]

If only Babbage could have worked for IBM! Despite all these resources, not least of all the incentive of the war effort, this machine, known as the ASCC or Mk. 1, started in 1937 and completed in 1944, was a very modest affair compared with the one that Babbage envisaged. Its store consisted of seventy-two counters each capable of holding twenty-three digits, compared with Babbage's thousand variables of fifty figures each. It had no device for reversing the flow of input so that an operation could not be repeated, and its output was so modest, that, at best, only tables of Bessel functions could be produced.

The first electronic computer, the Electronic Integrator and Calculator (ENIAC) was completed at the Moore School of Engineering, Philadelphia, in 1946. Its primary purpose was for projectile calculation but it could be adapted by a very complicated logical arrangement for performing other calculations such as the solution of partial differential equations. It is again an extraordinary reflection on this particular branch of technology that, although electronic techniques had been known since 1919, the need for such calculators had been manifest for the major part of the century and above all that Babbage had worked out all the major principles by 1842 at the latest, the first electronic computer was so slow in appearing. There is, unfortunately, no evidence that any of the early inventors of modern computers made use of Babbage's work, or were even aware of its existence. According to A. D. and K. H. V. Booth, 'The basic design of ENIAC lacks many of those characteristics foreseen by Babbage.'[54]

To recapitulate these characteristics foreseen by Babbage, the major elements are:

1. The five basic units of a computer consisting of:
   (a) the *store* containing the data, instructions and intermediate calculations;
   (b) the *mill* in which the basic arithmetical operations are performed;
   (c) the *control* of the whole operation in Babbage's case by means of a Jacquard loom system;
   (d) the *input* by means of punched cards;
   (e) the *output* which automatically prints results.

2. The registering apparatus by means of which instructions could be *repeated* as many times as desired.

3. The capability of the analytical engine to make *decisions* in the course of a calculation, and reverse the sequence of cards to continue the operation at any desired place in the programme.

4. The logically contrived *programme* to analyse any calculation into basic steps that the engine could perform in sequence.

5. The realisation that the engine was capable of performing any analytical calculation.

6. The amount of storage space required. It is remarkable, as Booth and Booth point out,[55] that when the ENIAC was under construction von Neumann and others devoted considerable thought to the optimum size of storage. They concluded that it was a capacity of 4096 words with six to twelve digits each. This is very comparable with Babbage's suggestion of 1000 numbers each of fifty digits.

All of these characteristics are included in modern computers, and apart from the obvious improvements resulting from advanced technology, they differ very little in principle from the analytical engine. It would be fitting to conclude this chapter by quoting the words of B. V. Bowden, when introducing his book on computing, *Faster than Thought* in 1953:

> Babbage's ideas have only been properly appreciated in the last ten years, but we now realize that he understood clearly all the fundamental principles which are embodied in modern digital computers. His contemporaries marvelled at him though few of them understood him; and it is only in the course of the last few

years that his dreams have come true. The rest of this book is devoted to an account of the construction and use of the machines which his vision inspired.[56]

NOTES

1. C. Babbage, *Passages from the Life of a Philosopher*, London, 1864, p. 42.
2. *Ibid.*, pp. 42–3.
3. *Memoirs of the Astronomical Society*, London, 1822, **1**, 309.
4. H. P. Babbage, *Babbage's Calculating Engines*, London, 1889, p. 213.
5. *Ibid.*, p. 214.
6. *Ibid.*, p. 212.
7. *Ibid.*, p. 215.
8. C. Babbage, *op. cit.*, p. 435.
9. *Ibid.*, p. 68.
10. Parliamentary Paper, No. 370, p. 1823.
11. British Museum Additional Manuscripts 37182, No. 425.
12. C. Babbage, *op. cit.*, p. 71.
13. F. Baily, *Edinburgh Journal of Science*, 1824, **1**, 141–3.
14. C. Babbage, *op. cit.*, p. 65.
15. H. P. Babbage, *op. cit.*, p. 232.
16. British Museum Additional Manuscripts 37184, No. 301.
17. *Ibid.*, No. 451.
18. *Ibid.*, No. 455.
19. *Ibid.*, No. 301.
20. D. Lardner, 'Babbage's calculating engines', *Edinburgh Review*, 1834, **59**, 263–327.
21. British Museum Additional Manuscripts 37192, No. 172.
22. *Ibid.*, No. 178.
23. C. Babbage, *op. cit.*, pp. 94–5.
24. British Museum Additional Manuscripts 37192, Nos. 189–194.
25. *Athenaeum*, 14 October, 1848.
26. *Philosophical Magazine*, 1843, **23**, 237.
27. *Journal of the Statistical Society*, 1871, **34**, 416.
28. C. Babbage, *op. cit.*, p. 98.
29. British Museum Additional Manuscripts 37195, No. 82.
30. C. Babbage, *op. cit.*, p. 103.
31. *Ibid.*, p. 107.
32. *Ibid.*, pp. 99–100.
33. *Ibid.*, pp. 113–14.
34. *Ibid.*, p. 112.
35. H. P. Babbage, *op. cit.*, p. 331.
36. C. Babbage, *op. cit.*, p. 117.
37. *Ibid.*, pp. 117–18.
38. *Ibid.*, p. 121.
39. *Ibid.*, p. 119.
40. *Ibid.*, p. 123.

41. *Ibid.*, p. 114.
42. H. P. Babbage, *op. cit.*, pp. 334–5.
43. C. Babbage, *op. cit.*, pp. 123–4.
44. *Ibid.*, p. 131.
45. *Ibid.*, p. 135.
46. *Ibid.*, p. 136.
47. *Scientific Memoirs*, 1843, **3**, 678.
48. C. Babbage, *op. cit.*, pp. 159–62.
49. *Napier Memorial Volume*, London, 1915, pp. 19–20.
50. C. Babbage, *op. cit.*, pp. 449–50.
51. H. P. Babbage, *op. cit.*, p. 323.
52. *Ibid.*, pp. 329–30.
53. L. J. Comrie, *Nature, London*, 1946, **158**, 567.
54. A. D. Booth and K. H. V. Booth, *Automatic Digital Calculators*, London, 1953, p. 17.
55. *Ibid.*, p. 18.
56. B. V. Bowden, *Faster than Thought*, London, 1953, p. 7.

9

Conclusion

It only remains now to summarise the mathematical work of Charles Babbage and to formulate an assessment of the importance of his contributions to the subject. We will discover the paradoxical conclusion that Babbage's mathematical achievements were quite considerable, but his immediate influence on mathematical thought and progress was almost negligible. It will be instructive to consider the reasons why this was so.

We have seen that Babbage lived through a period of extreme mediocrity in British mathematics. This meant that he had to acquire most of his knowledge by himself. Even at Cambridge, which was then the centre of any mathematical activity in this country, he quickly discovered that he could learn little from the formal tuition and had instead to devote himself to his own studies. He was fortunate at this time to have as undergraduate contemporaries, J. F. W. Herschel and George Peacock who were probably the only mathematicians in the country capable of working with him. They studied together the works of all the reputable Continental mathematicians, and, as the first chapter of the *Memoirs of the Analytical Society* indicates, their depth of reading and understanding was considerable by 1813. Although Babbage was strongly influenced by Continental rather than British mathematicians, and his work shows many instances of their ideas, his final output was entirely different from anything produced by his contemporaries. In fact, we find Babbage pursuing independent lines of research, working completely on his own apart from a little assistance from Herschel, with very few to teach or encourage or criticise him. These researches into pure mathematics which began in 1813 continued at a prolific rate, producing three books, two unpublished books, three papers of

220

considerable length, fourteen other papers, two long encyclopaedia articles, and then came to an abrupt end in 1821.

For the remaining fifty years of his life, Babbage worked hard
on the construction of his engines. He also published work on
magnetism, astronomy, life assurance, geology, biology,
economics, religion, politics, ciphering, submarines, ophthal-
moscopes, and machinery, but never anything else on pure
mathematics. It will again be a matter of interest to investigate
how a man of such creative mathematical talent could virtually
retire from the subject at the age of thirty having won the acclaim
not only of those of his own countrymen who could understand
but also of great foreign mathematicians like Cauchy.

What then were his major achievements during this intensive
eight-year period? We have observed already at some length the
calculus of functions, which he developed from the solution of a
few isolated problems into a systematic branch of mathematics,
solving a variety of very complicated examples and introducing
ingenious methods to develop the technique. He rightly regarded
this subject as the most general branch of analysis then existing,
and saw far more in the solution of functional equations than had
even been suspected by such mathematicians as D'Alembert,
Euler, Lagrange, Laplace and Monge.

In his unpublished work, 'The Philosophy of Analysis', at least
two major original ideas have been noted. The first concerned a
new approach to algebra which foresaw the subject liberated
from continued arithmetical interpretation and might have led
the way to the more modern formulations of the 1840s. In the
final chapter on the attempt to discover a mathematical interpretation of the intuitive logic required for playing noughts-and-
crosses, there is a remarkable evaluation of the successive values
of a random variable indicating a type of stochastic process.

Babbage's ideas on the importance of notation were fully
developed in various papers and he produced some very original
ideas for treating notation itself by systematic logical rules and
thus increase both the generality and the simplicity of mathematical reasoning. He also had thoughts on other branches of
mathematics, mainly geometry, probability, the theory of numbers, finite and infinite series and products.

In addition to these achievements in pure mathematics, we
must also consider Babbage's major contribution to science and
technology, the analytical engine, which I have explained was in

every way comparable to a modern computer. As by-products in his computer development, it might be said that Babbage invented systems analysis and operations research.

Now modern algebra, stochastic processes, computer science, operations research and systems analysis are considered to be very up-to-date branches of mathematics today in both universities and industry, so it might be thought that Babbage's discoveries were significant indeed. However, it is idle to attempt to fasten labels of invention of subjects onto Babbage or to try to regard him as a second Newton heralding a new era of mathematics. It has been said that divisions of subject are largely a mere academic administrative convenience and that a man of a highly individual degree of skill sees only a large number of problems to be solved by any logical means. Babbage certainly comes into this latter category. He tackled a whole variety of problems, drawn many times from areas unfamiliar to the mainstream of mathematics, and attempted to solve them by his own devices without ever being aware of creating a new 'subject'. If he unintentionally crossed paths not explored again for another hundred years, this indicates only that his mind was extraordinarily aligned to middle twentieth-century thinking, without his being in any way responsible for the later developments in mathematical science. As has been said already, his immediate influence on his contemporaries was almost negligible.

I will attempt to establish this last remark by looking at the fate of his major discoveries one by one: The calculus of functions, so admirably developed into a systematic study by Babbage, was not continued on the same lines by anyone, and the suggestions he made for future development have never been acted upon. Apart from enthusiastic comment by Bromhead in England and Cauchy in France, it seems that no prominent mathematician has studied the two essays published in the *Philosophical Transactions*, with the possible exception of G. Boole, who made a favourable reference to Babbage's work in his *Calculus of Finite Differences*, London, 1844. Boole's work on functional equations was, however, at a much lower level than Babbage's. It is true that functional equations have been studied recently in a systematic way, but a large part of Babbage's ideas has not been developed, particularly that concerning what he called periodic functions. S. Pincherle in 1912 was kind enough to describe the functional equation equivalent to $\psi^n x = x$ as 'l'équation de Babbage', but there is no evidence that more modern mathematicians have interested themselves in Babbage's equation. I would venture to suggest that with a properly rigorous approach, and a generalisation possibly through modern topological methods, there is great potential still to be drawn from the two essays on the calculus of functions.

Little need be said about the ideas from 'Philosophy of Analysis', for these were never published, and it is quite possible that no-one ever read them, apart from George Peacock. It is still interesting to note that several books relating to the history of mathematics, such as that by H. Eves and C. V. Newsom (*An Introduction to the Foundations and Fundamental Concepts of Mathematics*, New York, 1961), give Peacock the credit for stating the rules for a reformulation of algebra in more general terms in 1830, when we have seen that these rules were stated in almost identical fashion by Babbage in 1821. I believe that Peacock was too honourable a man for there to be any question of plagiarism, but the coincidence is extraordinary, particularly when we have his letter stating that he had studied these essays by Babbage. The fact that each made exactly five points with practically isomorphic content suggests a very high degree of subconscious assimilation by Peacock. The absence of any known further correspondence between the two on this subject is most annoying.

I have suggested in a fairly lighthearted way that Babbage invented systems analysis, because it seems that this claim can be made on grounds as good as those put forward by people who advocate that Babbage invented operations research. The evidence for this latter claim is the work which he did for the Post Office and the pin-making industry. Both of these tasks consisted of a cost analysis of the fundamental operations involved with subsequent recommendations about the principles on which prices should be estimated. To consider the Post Office case as typical and briefer, Babbage states in one of his 'concluding chapters' from *Passages from the Life of a Philosopher*:

> When my friend, the late General Colby, was preparing the materials and instruments for the intended Irish survey, he generally visited me about once a week to discuss and talk over with me his various plans. We had both of us turned our attention to the Post Office, and had both considered and advocated the question of a uniform rate of postage. The ground of that opinion was, that the actual *transport* of a letter

formed but a small item in the expense of transmitting it to its destination: whilst the heaviest part of the cost arose from the *collection* and *distribution*, and was, therefore, almost independent of the length of its journey.[1]

On this basis, that the cost of postage depended mainly on that of collection and distribution, and was almost independent of distance, the decision was made to charge a uniform rate for conveying a letter. The claim to originality by Babbage is in fact disputed by J. D. Hill,[2] who asserts that the invention of the standard rate principle is due solely to Sir Rowland Hill. I have to agree that there appears to be no evidence to support Babbage other than his own assertion in *Passages*. It is surprising that Babbage, whose inventiveness was so prolific, should lay claim to something he did not invent, but somehow or other, years after the event, he seems to have convinced himself. Even if he did make this discovery, there is in this example, as in that applied to the pin-making industry, an excellent analysis on a cost basis of operations performed, but no techniques other than common-sense are suggested for dealing with the data. If this is to be counted as the invention of operations research (before 1827, according to Babbage), then it can be equally claimed that the analysis in 1821 of the French method of compiling tables, with the suggestion that most of the human drudgery be replaced by a computer, be considered the invention of systems analysis. In both cases, Babbage's ideas exerted no influence on the modern methods used when first operations research began as a recognisable discipline in 1937 and then systems analysis more recently.

However, the most obvious example of Babbage's genius having been ignored arose from his invention of the analytical engine. It has been described in chapter 8, how this engine was in every way comparable to a modern computer, excepting only, of course, the electronic techniques. Babbage died in 1871 with the construction far from finished. His son, H. P. Babbage, completed small parts of the engine in accordance with the specification and found that they worked perfectly. The British Association, however, set up a committee in 1878 who recommended that no attempt should be made to complete the engine, and from this time onwards the project was forgotten.

The first fully automatic computer was not completed until 1944, and was, despite unlimited finance, modern techniques and the incentive of the war effort, a modest affair compared with the one that Babbage had envisaged. If the inventors had studied his work, their task would have been made more easy and fruitful. It is surprising that, in spite of the need for computers in the twentieth century, the knowledge of electronics since 1919, and the existence of Babbage's ideas for all the major principles by 1843, the first such machines were so slow to materialise. In passing, it should also be noted that the algebraic techniques for the logical design of computers were invented by Babbage's contemporaries, Boole and De Morgan.

It seems almost a general principle that Babbage's mathematical ideas were so little appreciated by his contemporaries that they exerted no subsequent influence on the development of mathematics, even though twentieth-century experience has demonstrated their major importance.

The influence that Babbage did exert in mathematics was by way of reform, organisation and suggestion rather than by any particular discovery.

At Cambridge he was undoubtedly the originator and the mainspring of the Analytical Society. His purpose was the reform of mathematics in the university, which effectually meant a reform in the country as a whole. It was largely as a result of his perseverance that the differential notation was introduced into Great Britain at a time when there was hostility between mathematicians here and on the Continent, at an irrational partisan level made worse by the Napoleonic wars. This reform was small in itself, but it carried with it the wider implication that British mathematicians were freed from restrictions in their work because of patriotic prejudice. British mathematics did not profit immediately, and it is interesting to note the coincidence that, when Babbage ceased publishing mathematical work in 1821, there were hardly any publications by any British mathematician for the next ten years; but the seed had been sown, and by 1840 a new generation of mathematicians had emerged, such as Boole, Hamilton, De Morgan, Maxwell, Cayley and Sylvester, producing work comparable to anything on the Continent.

Throughout his life, Babbage was concerned with the organisation of science, and the status of the scientist and mathematician. He was motivated generally by a sense of the mutual responsibility of science and society, and his unfortunate personal experiences added vitriol to his criticisms of the establishment. He wrote a book, *Observations on the Decline of Science in England*, London,

1830, in which he strongly attacked the way in which the Royal Society failed to give sufficient encouragement to research. At this time, it was far too easy to become a Fellow of this Society, and it was found that, out of a total of 630 Fellows, only about a hundred were genuine working scientists. In the same year, Babbage campaigned to have his friend J. F. W. Herschel elected as President, but was disappointed to see him narrowly defeated by a non-scientist, the Duke of Sussex.

At this time, Babbage travelled on the Continent a great deal, where he made the acquaintance of most of the leading scientists and wrote enthusiastic reports about their Conventions. As a result, he helped to form the British Association for the Advancement of Science in 1831. At the third meeting of the British Association at Cambridge in 1833, he suggested the idea of a Statistical Section which aroused immediate interest. It was soon found that the great amount of data available necessitated a more permanent society.

> The interest of our discussions, and the mass of materials which now began to open upon our view, naturally indicated the necessity of forming a more permanent society for their collection. The British Association approved of the appointment of a permanent committee of this section. I was requested to act as chairman, and Mr. Drinkwater as secretary. On the 15th March, 1834, at a public meeting held in London, the Marquis of Lansdowne in the chair, it was resolved to establish the Statistical Society of London.[3]

Earlier he had been very much involved in the formation of the Astronomical Society. In *The History of the Royal Astronomical Society 1820–1920*, edited by J. L. E. Dreyer,[4] it is reported that Babbage was one of fourteen men who met at the Freemasons' Tavern, Great Queen Street, Lincoln's Inn Fields, on 12 January 1820, and resolved to constitute the Astronomical Society of London. A committee of eight selected to draw up rules and regulations consisted of C. Babbage, F. Baily, T. Colby, H. T. Colebrooke, O. G. Gregory, J. F. W. Herschel, D. Moore and the Rev. W. Pearson.

While it is true that Babbage's specifically mathematical output ceased about 1821, and his later publications deal with other matters, there can be little doubt that his thinking remained primarily mathematical. Whether discussing miracles, life peerages, pin-making or public transport, the problems either receive mathematical treatment, or failing this, the mind of a mathematician is always evident. As H. W. Buxton said when summarising his life and qualities:

> Mr. Babbage was essentially a mathematician and regarded mathematics as the best preliminary preparation for all other branches of human knowledge; not even excepting theology, for he believed that the study of the works of nature with scientific precision, was a necessary and indispensable preparation to the understanding and interpreting their testimony of the wisdom and goodness of their Divine Author.[5]

Instead of asserting that Babbage's mathematical life ended at the age of about thirty, it can be seen that his interest in other matters may be interpreted rather as a sign of maturity, when he began to apply his mathematical training to the solution of problems in a non-mathematical world. We have seen that his hold on life as a child was extremely precarious, and as he gradually overcame his many illnesses and fears, he found security in the mathematical world, where he was master. His early researches show a primary interest in strictly pure mathematics as exemplified in his superb work on the calculus of functions. Then, as the young mathematician becomes more drawn into the affairs of the world, he encounters problems that do not fit into neat mathematical categories, and he finds that either he has to take the easier way of dividing his mind into logical and non-logical compartments, or to use his mathematical ability to help answer the less rationally framed questions. In view of the great difficulty of the latter course, most mathematicians choose the simpler solution, but Babbage's integrity led him to take his mathematics directly into the world. We find his later mathematical papers providing some of these problems, arising for example from economics and simple games, and then we have his most ambitious attempts to solve, still by basic mathematical reasoning, the type of problem in which not even one equation can be formulated. The success Babbage achieved in such a difficult task can be attributed to the preparation of his mind through his earlier years of preoccupation with pure mathematics. According to Maboth Moseley's biography:

> He made acute analyses of the pin-making industry and the printing trade. He analysed the economics of the Post Office, as a result of which Sir Rowland Hill introduced the penny post.

> He studied insurance records and published the first comprehensive treatise on actuarial theory and the first reliable life tables. He invented the heliograph and the ophthalmoscope . . . He took a lifelong interest in ciphers and deciphering, and he could pick any lock. He wrote a ballet, and invented coloured lighting for the stage.[6]

These achievements, which are by no means an exhaustive list of all that Babbage accomplished, belong to the latter fifty years of his life, and could all be said to be inspired by a mind that was essentially mathematical, and, as it matured, sought to bring this quality into the solution of non-mathematical problems.

Because Babbage's mind ranged over so many ideas, it is not surprising to find an unevenness in the power of his thought. In the course of this book it has been necessary several times to indicate weaknesses in Babbage's mathematical reasoning, not all of them resulting from the general paucity of mathematics in Britain at that time. His autobiography shows that his mind was capable of dealing with the extreme complexities of the analytical engine, and yet could be diverted to giving serious attention to utterly trivial matters from dealing with organ-grinders to methods of salesmanship. In either circumstance, Babbage was never dull, and even his more 'trivial thinking' can be attributed to the restless mental endeavours of a man of exceptional talent. He states several times in his mathematical writings his intention to display the workings of his mind in arriving at his inventions, without succeeding directly in conveying much information about this. We can, however, define the strength of his reasoning as an ability to relate problems from entirely different fields, to formulate a mathematical treatment from the data, and to generalise from particular cases.

It is not the purpose of this book to attempt a general assessment of Babbage's personality, but as several characteristics have arisen in the course of this work, particularly in the previous chapter, it is only fair to end with a wider assessment, which is contrary to many popular misconceptions about him.

Maboth Mosely has said:

> Contentious, contrary, sarcastic and cynical, there was another side to his character. The son of a rich London banker, he was no unapproachable scientific recluse hiding away in an ivory tower. He was a sociable, witty and loveable man, with a ringing hearty laugh and an unfailing fund of amusing anecdotes on

which he constantly dined out. Invitations to his Saturday-evening parties were eagerly sought, the indispensable qualifications being intellect, beauty or rank. His son records that some 200 to 300 people would attend one of these gatherings, and that on one occasion no fewer than seven bishops were present.[6]

A more contemporary assessment was given by H. W. Buxton, who said:

> It has even been thought, because his attention has been so long devoted to the consideration of abstract ideas and severe technical details, that he could not be susceptible to the more tender sensibilities of humanity. Nothing was more remote from the truth, and to the last period of his life he participated warmly in the most delicate and solid friendships; indeed, his sensibility was to the last unimpaired by time, and remained unabated by the thousand disturbing causes which always surrounded and distracted him.[7]

Perhaps the most enduring of Babbage's achievements will prove to be his demonstration of the intellectual value of manual work, his insistence on the proper status and dignity of the professional scientist, and his example of the value of mathematical reasoning in a non-mathematical world.

NOTES

1. C. Babbage, *Passages from the Life of a Philosopher*, London, 1864, p. 447.
2. J. D. Hill, 'Charles Babbage, Rowland Hill and Penny Postage', *Postal History International*, 1973, **2**(1).
3. C. Babbage, *op. cit.*, p. 434.
4. H. H. Turner, 'The decade 1820–1830'. In J. L. E. Dreyer (ed.), *The History of the Royal Astronomical Society 1820–1920*, London, 1923, pp. 1–49.
5. H. W. Buxton, 'Memoir of the Life and Labours of the late Charles Babbage Esq.' unpublished, p. 1986.
6. M. Moseley, *Irascible Genius*, London, 1964, pp. 20–1.
7. H. W. Buxton, *op. cit.*, pp. 2016–18.

APPENDIX
Mathematical books and papers by Charles Babbage

Published books

1 *Memoirs of the Analytical Society*, Cambridge, 1813. (Written with J. F. W. Herschel.)
2 Translation of S. F. Lacroix's *'Sur le Calcul Différentiel et Intégral'*, Cambridge, 1816. (Written with J. F. W. Herschel and G. Peacock.)
3 *Examples of the Solutions of Functional Equations*, Cambridge, 1820.

Unpublished books

1 'The History of the Origin and Progress of the Calculus of Functions during the years 1809, 1810 ... 1817' (Museum of the History of Science, Oxford).
2 'The Philosophy of Analysis', *c*. 1821 (British Museum).

Papers

1 'An essay towards the calculus of functions, Part I', *Philosophical Transactions*, 1815.
2 'An essay towards the calculus of functions, Part II', *Philosophical Transactions*, 1816.
3 'Demonstration of some of Dr. Matthew Stewart's general theorems, to which is added an account of some new properties of the circle', *Journal of Sciences and the Arts*, 1817.
4 'Observations on the analogy which subsists between the calculus of functions and other branches of analysis', *Philosophical Transactions*, 1817.
5 Solutions of some problems by means of the calculus of functions', *Journal of Science and the Arts*, London, 1817.
6 'Note respecting elimination', *Journal of Science and the Arts*, London, 1817.
7 'An account of Euler's method of solving a problem relating to the Knight's move at chess', *Journal of Science and the Arts*, 1817.
8 'On some new methods of investigating the sums of several classes of infinite series', *Philosophical Transactions*, 1819.
9 'Demonstration of a theorem relating to prime numbers.' *Edinburgh Philosophical Journal*, 1819.
10 'Observations on the notation employed in the calculus of functions', *Transactions of the Cambridge Philosophical Society*, 1821.

11 'An examination of some questions connected with games of chance', *Transactions of the Royal Society of Edinburgh*, 1823.
12 'On the application of analysis to the discovery of local theorems and porisms', *Transactions of the Royal Society of Edinburgh*, 1823.
13 'On the determination of the general terms of a new class of infinite series', *Transactions of the Cambridge Philosophical Society*, 1827.
14 'On the influence of signs in mathematical reasoning. *Transactions of the Cambridge Philosophical Society*, 1827.
15 'On notation', *Edinburgh Encyclopaedia*, 1830.
16 'On porisms', *Edinburgh Encyclopaedia*, 1830.

INDEX